ECE/EB.AIR/47

ECONOMIC COMMISSION FOR EUROPE
Geneva

AIR POLLUTION STUDIES 12

The State of Transboundary Air Pollution

*Report prepared within the framework of the
Convention on Long-range Transboundary Air Pollution*

UNITED NATIONS
New York and Geneva, 1996

NOTE

Symbols of United Nations documents are composed of capital letters combined with figures. Mention of such a symbol indicates a reference to a United Nations document.

*

*   *

The designations employed and the presentation of the material in this publication do not imply the expression of any opinion whatsoever on the part of the Secretariat of the United Nations concerning the legal status of any country, territory, city or area, or of its authorities, or concerning the delimitation of its frontiers or boundaries.

ECE/EB.AIR/47

UNITED NATIONS PUBLICATION

Sales No. E.96.II.E.21

ISBN 92-1-116653-5
ISSN 1014-4625

TABLE OF CONTENTS

Page

INTRODUCTION AND SUMMARY 1

Part One.   Strategies and Policies for
Air Pollution Abatement: 1995 Review

INTRODUCTION: MANDATE AND AIM OF THE REVIEW 3

I. EMISSION LEVELS AND TRENDS IN THE EFFECTS OF AIR POLLUTANTS 3

    A.   National annual total emissions in the ECE region, 1980-2010 3

    B.   National annual emissions by source category for the years 1985, 1990-
       1993 and 2000 4

II. NATIONAL STRATEGIES 4

    A.   General objectives and targets of air pollution abatement policy 4
       1. Basic principles 4
       2. General objectives and strategies 4
       3. Emission reduction targets 6

    B.   Legislative and regulatory framework, including national plans and
       programmes 7

    C.   Integrating air pollution policy and energy, transport and other policy areas 8

    D.   Administrative structures: national and local authorities 9

III. NATIONAL POLICY MEASURES 9

    A.   Regulatory provisions 9
       1. Ambient air quality standards 9
       2. Target loads or deposition standards 11
       3. Fuel quality standards 11
       4. Emission standards 12
       5. Licensing of potentially polluting activities 17
       6. Product-oriented requirements and labelling 18
       7. Other regulatory measures 18

    B.   Economic instruments 18
       1. Emission charges and taxes 18
       2. Product charges, taxes and tax differentiation, including fuel taxes 20
       3. User and administrative charges 21
       4. Emission trading 21
       5. Subsidies and other forms of financial assistance 21

    C.   Measures related to emission control technology 22
       1. Technology requirements in legislation and regulations 22
       2. Control technology requirements for stationary sources 22
       3. Control technology requirements for mobile sources 23
       4. The availability of unleaded fuel 24

    D.   Monitoring and assessment of air pollution effects 24
       1. Monitoring of air quality and environmental effects 24
       2. Research into air pollution effects and assessment of critical loads and
          levels 25

IV. INTERNATIONAL ACTIVITIES 26

    A.   The Convention 26

    B.   Activities aimed at improving the exchange of control technology 26

    C.   Other bilateral activities in the ECE region 26

V. Conclusions 27

    A. Status of implementation of the 1985 Helsinki Protocol on the Reduction of Sulphur Emissions or their Transboundary Fluxes by at least 30 per cent 27

    B. Status of implementation of the 1988 Sofia Protocol concerning the Control of Emissions of Nitrogen Oxides or their Transboundary Fluxes 27

Table 1. Emissions of sulphur (1980-2010) in the ECE region 29
Table 2. Emissions of nitrogen oxides (1980-2010) in the ECE region 31
Table 3. Emissions of ammonia (1980-2010) in the ECE region 33
Table 4. Emissions of non-methane volatile organic compounds (1980-2010) in the ECE region 35
Table 5. Emissions of methane (1980-2010) in the ECE region 37
Table 6. Emissions of carbon monoxide (1980-2010) in the ECE region 39
Table 7. Emissions of carbon dioxide (1980-2010) in the ECE region 41
Table 8. Convention on Long-range Transboundary Air Pollution and its related Protocols 43
Table 9. Emissions of sulphur (1980-2010) in the ECE region as a percentage of 1980 levels (and Figure I) 44
Table 10. Emissions of nitrogen oxides (1980-2010) in the ECE region as a percentage of 1987 levels (and Figure II) 46

Part Two. Forest Condition in Europe: 1994 Survey

Summary 49

Introduction 50

  I. Methods of the 1994 surveys 50

    A. Transnational survey 50

    B. National surveys 51

    C. Selection of sample trees 51

    D. Assessment parameters and data presentation 51

  II. Results of the 1994 surveys 52

    A. Transnational survey results 52
      1. General results 52
      2. Forest condition by species groups 53
      3. Defoliation and discoloration by mean age 53
      4. Easily identifiable damage 54
      5. Changes in defoliation and discoloration from 1993 to 1994 55
      6. Changes by climatic region 55
      7. Changes by species group 56
      8. Changes in defoliation since 1988 60

    B. National survey results 60

  III. Interpretation 61

  IV. Conclusions and recommendations 62

Table 1. Defoliation and discoloration classes according to ECE and EU classification 51
Table 2. Percentages of defoliation for broadleaves, conifers, and all species 52
Table 3. Percentages of discoloration for broadleaves, conifers and all species 52
Table 4. Percentages of defoliation of all species by mean age 53
Table 5. Percentages of discoloration of all species by mean age 54
Table 6. Percentages of trees with defoliation >25 per cent and discoloration >10 per cent by identified damage types, based on a total of 4,756 plots with 102,288 (defoliation) and 97,078 (discoloration) trees, respectively 55
Table 7. Percentages of the total tree sample and the common sample trees in different defoliation and discoloration classes in 1993 and 1994 56
Table 8. Changes in defoliation observed between 1993 and 1994 in classes 2-4 60
Figure 1. Percentages of defoliation of the common sample trees in 1993 and 1994 for each of the 10 climatic regions and for the total sample of CSTs 57
Figure 2. Percentages of discoloration of the common sample trees in 1993 and 1994 for each of the 10 climatic regions and for the total sample of CSTs 58
Figure 3. Development of defoliation for coniferous trees (defoliation classes 2-4) common to 1988-1994 59
Figure 4. Development of defoliation for broad-leaved trees (defoliation classes 2-4) common to 1988-1994 59

Annex I.   Forests and surveys in European countries (1994) 64
Annex II.   Defoliation of all species by classes and class aggregates (1994) 65
Annex III.   Defoliation of conifers by classes and class aggregates (1994) 66
Annex IV.   Defoliation of broadleaves by classes and class aggregates (1994) 67
Annex V.   Defoliation of all species (1986-1994) 68
Annex VI.   Defoliation of conifers (1986-1994) 69
Annex VII.   Defoliation of broadleaves (1986-1994) 70

Part Three.   Calculation of Critical Loads of Nitrogen as a Nutrient
Summary Report on the Development of a Library of Default Values

INTRODUCTION 71

   I.   OBJECTIVES AND METHOD OF WORK 71

  II.   DATA PRESENTATION 72

       A.   Data listing for each variable 72

       B.   Summary data 72

       C.   Mapping manual 72

 III.   PROGRESS MADE AND RESULTS ACHIEVED 72

       A.   Critical total annual leaching of nitrogen 72

       B.   Critical annual level of nitrogen immobilization 73

       C.   Critical annual removal of nitrogen 73

       D.   Critical annual flux of nitrogen to the atmosphere 73

       E.   Critical nitrogen losses in smoke from fires 73

       F.   Critical annual nitrogen losses through erosion 74

Part Four.   European Sulphur and Nitrogen Emissions, Depositions
for 1980 and 1993 and Export/Import Budgets

Table 1.   Letter codes of countries and other areas 76
Table 2.   Deposition of oxidized sulphur in 1980 77
Table 3.   Deposition of oxidized nitrogen in 1980 78
Table 4.   Export/import budgets of oxidized sulphur and nitrogen for 1980 79
Table 5.   Deposition of oxidized sulphur in 1993 80
Table 6.   Deposition of oxidized nitrogen in 1993 81
Table 7.   Deposition of reduced nitrogen in 1993 82
Table 8.   Export/import budgets of oxidized sulphur and nitrogen for 1993 83
Table 9.   Selected results from the time series analysis of monitoring results—Sulphur dioxide. Me-
              dian annual reductions, and concentration reductions from 1980 to 1993 in per cent of the
              1980/81 concentrations 84
Table 10.   Selected results from the time series analysis of monitoring results—Sulphate in particles.
              Median annual reduction, and concentration reduction from 1980 to 1993 in per cent of the
              1980/81 concentrations 84

Figure 1.   Total deposition of oxidized sulphur in 1980 85
Figure 2.   Total deposition of oxidized sulphur in 1990 86
Figure 3.   Change in exceedance of the critical sulphur deposition 1980-1990 87

INTRODUCTION AND SUMMARY

This twelfth volume of the series of *Air Pollution Studies*, published under the auspices of the Executive Body for the Convention on Long-range Transboundary Air Pollution, contains the documents reviewed and approved for publication at the thirteenth session of the Executive Body held at Geneva from 28 November to 1 December 1995.

Part One is the Annual Review of Strategies and Policies for Air Pollution Abatement. It updates the 1994 Major Review (*Strategies and Policies for Air Pollution Abatement*, Sales No. E.95.II.E.15) on the basis of national data and reports received up to February 1996. The focus of the review is on policy measures to implement the 1979 Convention and its related protocols on sulphur compounds (Helsinki, 1985), nitrogen oxides (Sofia, 1988) and volatile organic compounds (Geneva, 1991). Recent legislative and regulatory developments are summarized country by country, including ambient air quality standards, fuel quality standards and emission standards, as well as economic instruments for air pollution abatement. Information is provided on control technology requirements in operation in Europe and North America for stationary and mobile emission sources, and on existing administrative structures for air quality management, monitoring, assessment and research. In addition to ongoing activities under the Convention, the review also covers other multilateral and bilateral arrangements in the ECE region and related global concerns. The tables show national emission data and forecasts for sulphur dioxide (SO_2), nitrogen oxides (NO_x), volatile organic compounds (VOCs), ammonia (NH_3) and carbon dioxide (CO_2) from 1980 to 2010. Conclusions are drawn concerning the status of implementation of the sulphur and nitrogen oxides protocols on the basis of these data.

Part Two is an executive summary of the 1994 Report on the Forest Condition in Europe (*Forest Condition in Europe: Executive report on the results of the 1994 Survey*. 31 pp.). The main objective of this report is to give a condensed description of the condition of forests in Europe, as it has been assessed by the transnational and national annual surveys, carried out jointly by ECE under the Convention on Long-range Transboundary Air Pollution and by the European Community (EC). The survey presents results from 32 European countries, referring to some 30,000 sample plots with about 648,000 sample trees. Some 178 million hectares have been covered by the surveys conducted in accordance with common guidelines. The results of the 1994 survey indicate that a significant proportion of forests in Europe shows signs of defoliation and/or discoloration. Data for different coniferous and broad-leaved species reflect the health of forests at the species level and also indicate which countries and age groups are the most affected. In most cases no evident source of damage was identified. Nevertheless, some countries regard air pollution as the essential factor causing forest damage in their countries. Most other countries consider air pollution as a factor leading to the weakening of forest ecosystems.

Part Three is a summary report on the development of a library of default values for each of the input variables to the simple mass balance equation for the calculation of critical loads of nitrogen and for a range of ecosystems. Part Four presents the modelling results of European sulphur and nitrogen emissions, depositions for 1980 and 1993, and export/import budgets.

For the presentation of these reports, some necessary editing has been done, but care has been taken to avoid any substantive change in the documents as reviewed by the Executive Body for the Convention[1]. Sole responsibility for the text rests with the secretariat of the United Nations Economic Commission for Europe.

For further information, please contact :
United Nations
Economic Commission for Europe
Environment and Human Settlements Division
Air Pollution Section
Palais des Nations
CH-1211 Geneva 10
Fax: 41 22 917 0107

---

[1] Part One appeared as a document under the symbol EB.AIR/R.92; Part Two as EB.AIR/WG.1/R.109; and Part Three as EB.AIR/WG.1/R.108. Part Four is drawn from annex I to document EB.AIR/GE.1/26.

Part ONE

STRATEGIES AND POLICIES FOR AIR POLLUTION ABATEMENT: 1995 REVIEW

INTRODUCTION

Mandate and aim of the review

By the terms of the Convention on Long-range Transboundary Air Pollution, the Contracting Parties shall, *inter alia*, "endeavour to limit and, as far as possible, gradually reduce and prevent air pollution including long-range transboundary air pollution" (article 2); "develop . . . policies and strategies which shall serve as a means of combating the discharge of air pollutants" (article 3); and "exchange information on and review their policies . . . aimed at combating . . . the discharge of air pollutants" (article 4). Furthermore, by the terms of article 8 (*g*), they shall "exchange available information on national, subregional and regional policies and strategies for the control of sulphur compounds and other major air pollutants". The resolution on long-range transboundary air pollution adopted in 1979 at the High-level Meeting within the Framework of the ECE on the Protection of the Environment states that the Signatories to the Convention will seek to bring closer together their policies and strategies for combating air pollution including long-range transboundary air pollution (ECE/HLM.1/2, annex II).

The Helsinki Protocol on the Reduction of Sulphur Emissions or their Transboundary Fluxes by at least 30 per cent, calls upon Parties to develop "national programmes, policies and strategies . . . and report thereon as well as on progress towards achieving the goal to the Executive Body" (article 6). Finally, by the terms of article 8 (Information exchange and annual reporting) of the Sofia Protocol concerning the Control of Emissions of Nitrogen Oxides or their Transboundary Fluxes:

1. The Parties shall exchange information by notifying the Executive Body of the national programmes, policies and strategies that they develop in accordance with article 7 and by reporting to it annually on progress achieved under, and any changes to, those programmes, policies and strategies, and in particular on:

(*a*) The levels of national annual emissions of nitrogen oxides and the basis upon which they have been calculated;

(*b*) Progress in applying national emission standards required . . .;

(*c*) Progress in introducing the pollution control measures . . .;

(*d*) Progress in making unleaded fuel available;

(*e*) Measures taken to facilitate the exchange of technology; and

(*f*) Progress in establishing critical loads.

2. Such information shall, as far as possible, be submitted in accordance with a uniform reporting framework.

Similar provisions are made in the Geneva Protocol concerning the Control of Emissions of Volatile Organic Compounds or their Transboundary Fluxes, which was adopted and signed in 1991, and in the Oslo Protocol on Further Reduction of Sulphur Emissions, which was adopted and signed in 1994. These two protocols have, however, not yet entered into force.

This 1995 annual review updates information published in the 1994 Major Review on strategies and policies for air pollution abatement (ECE/EB.AIR/44) and should be read in conjunction with that review. It includes only information on recent developments reported since the 1994 Major Review. Only for Parties that had not submitted reports for the Major Review, is a full overview of their strategies and policies included. Furthermore, the information on national total emissions, their projections and emissions by source category have been included in full to allow an evaluation of trends.

In a letter sent out on 10 March 1995, Parties and Signatories were requested to submit the information required. To facilitate responses a questionnaire prepared on the basis of the annotated outline of the major review, EB.AIR/R.77, as amended by the Executive Body (ECE/EB.AIR/36, para. 18) was provided. For this annual review the secretariat received information from 33 Parties (Austria, Belarus, Belgium, Bulgaria, Canada, Croatia, Cyprus, Czech Republic, Denmark, France, Germany, Greece, Hungary, Iceland, Ireland, Italy, Liechtenstein, Luxembourg, Netherlands, Norway, Poland, Portugal, Russian Federation, Slovakia, Slovenia, Spain, Sweden, Switzerland, Turkey, Ukraine, United Kingdom, United States, and European Community) that forms the basis for this review.

I. EMISSION LEVELS AND TRENDS IN THE EFFECTS OF AIR POLLUTANTS

A. National annual total emissions in the ECE region, 1980-2010

The guidelines for estimation of emission data adopted by the Executive Body (EB.AIR/GE.1/R.65) call for the reporting on six substances; as requested by the Executive Body (ECE/EB.AIR/36, para. 15), Parties were in addition invited to submit data on carbon dioxide emissions as prepared and approved under the United Nations Framework Convention on Climate Change (UNFCCC) for information only.

Country-by-country data on total national emissions and current reduction plans for future years are presented in seven tables: table 1, sulphur dioxide emissions 1980-2010; table 2, nitrogen oxide emissions 1980-2010; table 3, ammonia emissions 1980-2010; table 4, emissions of non-methane volatile organic compounds 1980-2010; table 5, emissions of methane 1980-2010; table 6, carbon monoxide emissions 1980-2010; and table 7, carbon dioxide emissions 1980-2010. Where no data have been received from Parties in the context of this review, previously available data have been included. Data on sulphur emissions for the years 2000, 2005 and 2010 for Parties that signed the Protocol on Further Reduction of Sulphur Emission in June 1994 were taken from annex II to that Protocol (ECE/EB.AIR/40), except for Parties that reported current reduction plans that go beyond the Protocol obligations.

To estimate emissions, Parties generally employed source statistics and emission factors as defined within the CORINAIR programme for the 1990 inventory. In some cases, such as Italy, the emission data for the years 1985 to 1989 were estimated on the basis of the CORINAIR'85 methodology whereas estimates for subsequent years are based on the CORINAIR'90 methodology. Future recalculations based on the latest methodologies may lead to some adjustments. The final results of the CORINAIR'90 programme have not yet been included.

Emission projections were reported for the following two scenarios: (1) current reduction plans, reflecting the politically determined intention to reach specific targets; and (2) the baseline scenario, reflecting the state of legal/regulatory provisions in place by 31 December 1994 and, in addition, as an indication of the uncertainty of projections, an upper and a lower limit for emission projections. Data falling under category (1) are included in tables 1 to 7, whereas those falling under category (2) were not included in this document. However, they can be found in tables 8.1 to 8.6, projections of emissions of major air pollutants for the years 2000, 2005 and 2010, of document EB.AIR/R.92/Add.1 and can be obtained from the UN/ECE secretariat.

B.　National annual emissions by source category for the years 1985, 1990-1993 and 2000

National annual emissions of major air pollutants by source category are presented in tables 9.1-9.6 of document EB.AIR/R.92/Add.2, with a separate table for each pollutant and each Party for which data have been received. The source category split adopted with the guidelines is based on engineering principles consistent with national practice. Definitions of these source categories are taken from the report of the EMEP Workshop on Emission Inventory Techniques, Regensburg,

Germany, 2-5 July 1991. Items 1 to 10 comprise anthropogenic emissions, whereas item 11 is foreseen for biogenic and natural emissions uncontrolled by man. Emission inventory data by source category were requested for 1985 and every year from 1990 onwards; projections in this format were requested for the year 2000. This information can be obtained from the UN/ECE secretariat.

II.　NATIONAL STRATEGIES

A.　General objectives and targets of air pollution abatement policy

1.　Basic principles

A number of basic principles guide national policies for air pollution abatement in the ECE region:

(a) *Sustainability* or sustainable development is a long-term objective for the policies of many Parties. It was particularly stressed in the reports submitted by Austria, the Czech Republic, France, Ireland, the Netherlands, Portugal, the Russian Federation, and Slovenia;

(b) The *precautionary principle* in some form or other is applied as a guiding principle in Austria, Canada, Croatia, the Czech Republic, France, Germany, Ireland, Slovenia, Sweden, and Switzerland;

(c) The application of the *polluter-pays principle* at the national level is guiding national policy developments in Bulgaria, Croatia, Cyprus, the Czech Republic, France, Germany, Ireland, the Netherlands, Slovenia, Sweden, Switzerland, Turkey, and the United Kingdom;

(d) A number of Parties (Cyprus, Czech Republic, France, the Netherlands) set the objective for air pollution abatement policies to reduce emissions at their source, following a *pollution prevention* approach;

(e) In addition, Sweden applies a *substitution principle* in environmental legislation which stipulates that substances harmful to health and the environment should be replaced by less harmful ones.

2.　General objectives and strategies

Many Parties have developed action plans or long-term programmes to implement their strategies. Some Parties have specified objectives for air pollution abatement policies on the basis of the effects of those pollutants. Many Parties base their air pollution abatement policy on some notion of best available technology (BAT) or best available techniques not entailing excessive cost (BATNEEC). Most Parties, however, apply a combination of both source- and effect-oriented principles. The following paragraphs reflect the recent developments and changes concerning the basis for national strategies and their main objectives with respect to the 1994 Major Review.

(a) *Belgium*: As additional general objective and within the framework of the 1990 Third International Conference on the Protection of the North Sea, Belgium agreed to achieve a significant reduction (of 50 per cent or more) in atmospheric emissions by 1995, or by 1999 at the latest, of the substances specified in Annex 1A to the Ministerial Declaration of the Conference, provided that the application of best available technology, including the use of strict emissions standards, enables such a reduction. For substances that cause a major threat to the marine environment, and at least for dioxins, mercury, cadmium and lead, it agreed to achieve reductions of 70 per cent or more between 1985 and 1995 of total inputs (via all pathways), provided that the use of BAT or other low-waste technology enables such reductions.

(b) *Bulgaria*: The National Environment Strategy for 1991-2000 was adopted in January 1992 and updated in 1994. Two long-term programmes—on the reduction of sulphur and nitrogen oxide emissions and on the reduction of emissions of greenhouse gases by 2010—are being developed. They are expected to be completed in 1996.

(c) *Canada*: Two new policies that provide a framework for managing air pollution have been adopted in 1995. The Toxic Substances Management Policy calls for the elimination from the environment of toxic substances that result from human activity and are persistent and bioaccumulative. It also calls for the cradle-to-grave management of all other substances of concern that are released to the environment. "Pollution Prevention—A Federal Strategy for Action" is an approach to meeting the goal of shifting the focus from cleaning up pollution to preventing it in the first place.

(d) *Croatia*: The Law on Environmental Protection was passed in November 1994. It sets out the principle of prevention in environmental protection covering not only industrial and spatial policy, but also all other plans and programmes. With a view to implementing a systematic environmental policy, a national strategy is under preparation to serve as the principal basis for coordinating economic interests and overall development efforts with the need for environmental protection.

(e) *France*: France's strategy is based essentially on an emission reduction approach, through the application of the best available technologies at an economically acceptable cost. The regulations governing stationary sources are designed to promote the application of pollution and nuisance abatement technologies, under an integrated approach, aiming both to minimize overall pollution and to take account of the impacts of these types of pollution on the various environments. A bill on air quality which is to be put before Parliament at the beginning of 1996 is expected to propose a comprehensive and consistent approach embracing all the policies for air pollution control, based on the search for modes of sustainable development and the precautionary principle as well as the principles of prevention and public involvement and the polluter-pays principle. The forthcoming adoption of the Framework Directive on air quality by the Council of the European Union, together with the future law on air quality, should make it possible to elaborate on the "environment" approach of these poli-cies, by taking into account all the effects of air pollution (notably on buildings and on plants). This strategy will be refined and formalized when the air quality bill is discussed.

(f) *Hungary*: The national strategy on air pollution abatement focuses on: abatement in heavily polluted areas, such as the capital and some industrialized areas, especially aiming at improving the ambient air quality; maintaining the air quality in relatively "clean" regions; and fulfilling the obligations of international agreements on sulphur, nitrogen oxides, VOCs and CFCs. The main measures deriving from these objectives are: overall emission reductions; regional programmes to reduce emissions; emission control for the rapidly growing car fleet; development of the legal framework; and implementation of economic incentives and disincentives.

(g) *Liechtenstein*: The main objectives of air pollution abatement policy are: to protect humans, plants and animals and their communities against harmful and disturbing air pollution effects; to comply with immission limit values; and to reduce air polluting substances.

(h) *Russian Federation*: In the process of drawing up the national programme establishing priority areas and indicators for the reduction of emissions, a critical load map has been prepared for the European part of the Russian Federation on the basis of the EMEP grid. On the basis of this map, through modelling of different scenarios, the permissible levels of emission have been calculated for individual areas, due account being taken of commitments under the Convention and its protocols.

(i) *Slovakia*: In November 1993, the Strategy, Principles and Priorities of the State Governmental Environmental Policy was adopted. This document was based on an assessment of the environmental situation and includes objectives for international cooperation.

(j) *Slovenia*: Environmental objectives focus on sectoral policies, in particular those concerning energy, transport and agriculture. It is expected that the National Programme for Environment, which is to contain a synthesis of environmental policies in sectoral programmes, will be adopted in 1995.

(k) *Turkey*: The principles for environmental protection are laid down in the Sixth Five-Year Development Plan. The basic principle is to ensure the management of the environment so that human health and the natural equilibrium are protected, that the environment contributes to economic development and can still meet the needs of future generations. Probable environmental deterioration is to be forecast in advance and preventive measures against pollution will be taken. Measures will be initiated to improve the quality of petroleum products in order to prevent air pollution resulting from their consumption. In determining environmental standards, existing applicable technologies will be considered along with local conditions. Work has commenced to harmonize national legislation with European Community (EC) environmental policies.

3. Emission reduction targets

Most Parties specify abatement targets for sulphur, nitrogen oxides and volatile organic compounds. These are often more stringent than the emission reduction targets specified under the protocols. The emission figures for the years 1995 and onwards given in tables 1 to 7 provide an overview of current national reduction plans, i.e. politically determined emission reduction targets. The following subparagraphs present the targets derived from the protocols to the Convention and specified in national legislation:

(a) The Parties to the 1985 Protocol on the Reduction of Sulphur Emissions or their Transboundary Fluxes by at least 30 per cent, listed in table 8, undertook to reduce sulphur emissions by at least 30 per cent below 1980 levels by the end of 1993 and, as decided by the Executive Body, these levels should not be surpassed thereafter. The 1994 Protocol on Further Reduction of Sulphur Emissions specifies in its annex II a set of differentiated abatement targets for Parties. These have been included as current reduction plans in table 1 except for those Parties that reported stricter targets. Under this Protocol, *Canada* has established a Sulphur Oxides Management Area (SOMA) in south-eastern Canada, where an emission ceiling of 1750 kt p.a. as of the year 2000 applies. Furthermore, *Cyprus* intends to stabilize emissions from the year 2005 onwards. The *Netherlands* plans to reduce sulphur emissions to 92,000 tons by 2000 and to 56,000 tons by 2010. In *Poland*, a programme to limit sulphur dioxide emissions was adopted in 1988. Its targets are to stabilize emissions at 1980 levels by 1995, to reduce them by 30 per cent by the year 2000 and by 50 per cent by 2010. These targets were confirmed in 1991 and supplemented by a long-term target of an 80 per cent reduction to be achieved possibly around 2020. The stricter emission ceilings (up to 2010) of the 1994 Oslo Protocol, which Poland has signed, will replace the earlier targets. In *Slovenia*, the target, included in the Strategy for Energy Use, is to reduce by 2005 annual SO_2 emissions to a level below 30 kg per capita. The target in *Switzerland* is to bring sulphur dioxide emissions down to 1950 levels (i.e. a 57 per cent reduction compared to 1980 levels). The *United States* aims at cutting annual emissions of sulphur dioxide by 10 million tons below 1980 levels.

(b) The Parties to the 1988 Protocol concerning the Control of Emissions of Nitrogen Oxides or their Transboundary Fluxes, listed in table 8, have the obligation to stabilize nitrogen oxide emissions by the end of 1994 at 1987 levels (or 1978 for the United States). In addition, twelve Parties (Austria, Belgium, Denmark, Finland, France, Germany, Italy, Liechtenstein, Netherlands, Norway, Sweden and Switzerland) made a declaration to the effect that they will aim for a reduction of nitrogen oxide emissions in the order of 30 per cent based on emission levels of any year between 1980 and 1986 as soon as possible and at the latest by 1998. Furthermore, *Austria* aims at reducing its NO_x emissions by 40 per cent by the end of 1996, 60 per cent by the end of 2001 and 70 per cent by the end of 2006 below its 1985 levels. *Cyprus* intends to stabilize its emissions from the year 2005 onwards. In *Denmark*, based on an action plan, a 50 per

cent reduction in nitrogen emission from the agricultural sector into the aquatic environment and to the air is envisaged by the year 2000 based on 1985 levels, while NO_x emissions from the transport sector are to be reduced by 40 per cent before 2000 and 60 per cent before 2010, based on 1988 levels; total NO_x emissions are to be reduced by 35 per cent by the year 2000. The *Netherlands* plans to reduce nitrogen oxide emissions to 249,000 tons by 2000 and to 120,000 tons by 2010; it targets a maximum emission level for ammonia of 82,000 tons for 2000 and 50,000 tons for 2010. In *Poland* a programme to limit nitrogen oxide emissions was adopted in 1989 and confirmed in 1991; it calls for the stabilization of emissions at 1987 levels by 1994, a 10 per cent reduction by 2000 and a 50 per cent reduction by 2010. In *Sweden* regulations on ammonia emissions are planned; these would lead to approximately a 25 per cent decrease in national emission levels based on 1980 levels; a new target of a 50 per cent decrease is now under examination. The target in *Switzerland* is to reduce nitrogen oxide emissions to 1960 levels (i.e. a 69 per cent reduction compared to 1984 levels).

(c) The 1991 Protocol concerning the Control of Emissions of Volatile Organic Compounds or their Transboundary Fluxes specifies three options for emission reduction targets that have to be chosen upon signature:

(i) 30 per cent reduction in emissions of volatile organic compounds (VOCs) by 1999 using a year between 1984 and 1990 as a basis. (This option has been chosen by Austria, Belgium, Finland, France, Germany, Netherlands, Portugal, Spain, Sweden and the United Kingdom with 1988 as base year, by Denmark with 1985, by Italy and Luxembourg with 1990 and by Liechtenstein, Switzerland and the United States with 1984 as base year);

(ii) The same reduction as for (i) within a Tropospheric Ozone Management Area (TOMA) specified in annex I to the Protocol and ensuring that by 1999 total national emissions do not exceed 1988 levels. (Annex I specifies TOMAs in Norway (base year 1989) and Canada (base year 1988));

(iii) Finally, where emissions in 1988 did not exceed certain specified levels, Parties may opt for a stabilization at that level of emission by 1999. (This has been chosen by Bulgaria, Greece, and Hungary).

In addition, *Austria* aims at reducing its VOC emissions by 40 per cent by the end of 1996, 60 per cent by the end of 2001 and 70 per cent by the end of 2006 below its 1988 levels. In *Denmark* the target is to cut hydrocarbon emissions from the transport sector by 40 per cent before 2000 and by 60 per cent before 2010; total non-methane volatile organic compound (NMVOC) emissions are to be reduced by 30 per cent before the year 2000 with 1985 as the reference year. The *Netherlands* plans to bring VOC emissions down to 193,000 tons by 2000 and to 117,000 tons by 2010. *Sweden* has laid down a national target to cut VOC emissions to 50 per cent by the year 2000 based on 1988 levels. The target in *Switzerland* is to reduce emissions to 1960 levels (i.e. a 57 per cent reduction compared to 1984 levels).

B. Legislative and regulatory framework, including national plans and programmes

This section gives an overview of recent changes since the 1994 Major Review in the legislative and regulatory framework, while the contents of regulatory provisions are summarized in chapter III. Many Parties have framework legislation setting the basis for all environmental regulations or all those related to air pollution abatement. In other cases this legislation is scattered over a larger number of laws, sometimes even extending to the provincial level. The following paragraphs also summarize relevant legislation that is under preparation.

(a) *Austria*: The legislation under preparation includes: an ordinance on the phase-out of HCFCs (to be banned by the end of 2002); an amendment to the ordinance concerning bans and restrictions on the use of organic solvents; and an ordinance on emissions from kraft pulp pumps (sulphate processing).

(b) *Bulgaria*: A Clean Ambient Air Bill has been drafted and now awaits Parliament's approval. It will replace subordinate legislation concerning, *inter alia*, emission standards, licensing procedures, emission limits for large combustion plants, and warning levels.

(c) *Croatia*: The new Air Protection Law was passed in June 1995. It is based on the environmental protection legislation of the European Community and other international provisions, as well as the experience of industrialized countries. Subordinate legislation, such as regulations on ambient air quality and emission limit values for all sources, is under preparation and expected to be finalized by the end of 1995.

(d) *Czech Republic*: On 23 August 1995 the Government approved the State Environmental Protection Policy.

(e) *France*: The present framework is based essentially on two pieces of legislation: a 1961 law and a 1976 law on designated installations for environmental protection, which requires a large number of installations to obtain a permit and comply with specific rules to prevent all types of nuisance and pollution. The 1961 law serves as a basis for all the technical regulations governing the rational use of energy and sources other than fixed installations. However, its purely technical approach to emission sources, as well as the rigidity of the regulatory instruments used in applying it, make it impossible to pursue a consistent and comprehensive ''impacts'' approach or to ensure prevention by all the means necessary to combat emissions at their source, particularly as regards energy consumption and justifications for journeys. This is one of the fundamental reasons why the French Government is reviewing the regulatory machinery in this area. The new law should in particular make it possible to transpose into French law certain new obligations stemming from the framework directive on air quality and the IPPC directive, so that a comprehensive approach can be evolved based on sources and the impacts of efforts to combat atmospheric pollution.

(f) *Hungary*: New legislation to be issued in 1995 will in principle be harmonized with regulations accepted at the international level, taking as guiding basis legislation in industrialized countries, EC regulations and guidelines of the World Health Organization (WHO). The obligations entered into under international agreements will be implemented within the new Act on Environmental Protection. In addition, a number of legislative modifications are planned for 1995, including a modification of the zoning system, resulting in a change of ambient air quality levels.

(g) *Ireland*: An air quality management plan for Dublin is currently being prepared, the object of which is to protect and improve air quality in the capital. An air pollution dispersion model will be used as a tool in the implementation of the plan.

(h) *Liechtenstein*: The basis for air pollution legislation is the Law on the Abatement of Air Pollution and its ordinances. A master plan of action for air pollution abatement is under preparation.

(i) *Norway*: An intergovernmental working group has proposed regulations based on the Pollution Control Act setting mandatory limit values for NO_2, SO_2, PM_{10} and lead. The regulations are expected to enter into force during 1996. They will bring regulations into conformity with EC directives on air quality; in some cases limit values will be even more stringent than at EC level. The ozone directive of the EC will also be implemented as of 1995.

(j) *Slovakia*: A Government ordinance on emission standards, national ambient air quality standards, polluting substance and a categorization of sources of pollution is under preparation. Also under preparation are two ministerial decrees: one dealing with measurements of emissions and ambient air quality, the other introducing a programme for emission reductions.

(k) *Turkey*: The Environmental Law, which came into force in 1983 and endorses the polluter-pays principle, provides a framework for environmental legislation. The 1986 Air Quality Control Regulation was introduced to implement the Law. Its objective is to regulate atmospheric emissions of soot, smoke, steam and aerosols, to protect human beings and their environment from hazardous air pollution, to eliminate significant adverse environmental effects of air pollution and to ensure their prevention. The regulation specifies facilities which have significant negative effects on human health and the environment and regulates the granting of emission licences for their operation. Included among these facilities are power plants, and emission limit values for sulphur dioxide, nitrogen oxides and dust are defined for large combustion plants. The regulation also contains ambient air quality standards and fuel quality standards.

(l) *Ukraine*: Air pollution is controlled by Ukraine's Ambient Air Protection Act. The country now has a number of normative instruments setting out standards and rules for ambient air protection: the 1992 Procedure for the Setting and Levying of Charges for Environmental Pollution; the Regulations concerning the Procedure for Issuance of Permits for Special Use of Natural Resources and the Regulations concerning the Procedure for Establishing Limits on the Use of Natural Resources of Republican Significance (both approved in 1992).

C. Integrating air pollution policy and energy, transport and other policy areas

An increased integration of decision-making in some key policy areas that determine the level of pollution, such as transport, energy, agriculture, trade and economics, can be considered as a means of strengthening preventive measures and as complementing end-of-pipe control measures. The following subparagraphs, updating the information published in the 1994 Major Review, summarize the most important changes in these policy areas.

(a) *Bulgaria*: In 1993 an Energy Charter containing a long-term programme until 2010 was drawn up. It targets heat and electricity generation to reduce their environmental impact. The programme was updated in 1994/95 and is about to be adopted by the Government. Its key elements are: an increase in the share of hydroelectric power and renewable energy sources (solar and geothermal); replacing solid fuels by gas and lignite coal by local low-sulphur or imported coal in some power stations; launching a new nuclear power facility after 2005; and replacing steam generators with fluidized bed boilers or installing desulphurization equipment in some thermal power plants.

(b) *Czech Republic*: Environmental requirements are part of the energy policy. Flue gas desulphurization facilities are under construction at all large thermal power plants which will be in operation after 1998. Contracts have been concluded on sulphur removal for six major power plants of České Energetické Zavody Inc. with a total installed output of more than 4000 MW. In addition, fluidized-bed combustion boilers are under construction or being prepared and will include pressure and co-generation units for the production of electricity and heat based on the combined cycle. Similarly, completion of construction work at the Temelin nuclear power plant will facilitate the reduction of emissions through a decrease in the combustion of coal for power production.

(c) *Denmark*: In view of the need for a re-evaluation of the developments in energy use and resulting emissions in the period up until 2005, new proposals will be drawn up to ensure the goal of a 20 per cent reduction in carbon dioxide emissions based on 1988 levels. The Danish Energy Agency will carry out the analysis which will form the basis for the new energy plan to be finalized in 1995.

(d) *France*: As mentioned above, these areas have already been integrated in part, in so far as technical regulations relating to the use of energy and transport—vehicles in particular—already take into account the objectives of reducing air pollution. However, one of the aims of the air quality bill is to ensure that these air pollution control activities are integrated more broadly into energy policies and transport policies, in particular through better incorporation of these concerns at an earlier stage.

(e) *Ireland*: The transport strategy recommended in the Dublin Transportation Initiative (DFI) final report will provide the general policy framework for the future development of the transport system in the Greater Dublin Area. As a result of this the Dublin Transportation Office has been created. It will provide essential support to the various agencies, including: coordinating and monitoring the implementation of the DFI strategy; reviewing existing or proposed activities of the implementation agencies in relation to the DFI; reviewing and updating the DFI strategy at least once every five years; and undertaking or promoting public consultation, public awareness and education campaigns undertaken by the various agencies in relation to the DFI strategy.

(f) *Liechtenstein*: Besides existing legislation on the promotion of public transport and of energy conservation in buildings and plants, a number of new laws and ordinances to promote energy conservation are under preparation. They cover: a reconstruction programme for old buildings; stricter heat transmission requirements for new buildings; stricter energy regulations in the construction and building sector; and the promotion of renewable energies, the use of biogenic fuels and the replacement of fossil energy sources. In addition, a master plan for a traffic concept will be drawn up containing physical planning and space utilization regulations and promoting non-motorized traffic systems and public transport.

(g) *Netherlands*: A policy document on air transport and air pollution was prepared and issued to Parliament. It seeks to identify and take stock of the problem of air pollution due to air traffic, thus laying the basis for a consensus on measures to mitigate the harmful effects of aircraft emissions. The document lists possible measures to combat adverse environmental effects of these emissions. Their implementation will depend, *inter alia*, on their cost-effectiveness, safety and the economic consequences of implementation. A project group representing various employers' organizations in the logistics and transport sectors and three ministries is currently addressing the problem of nitrogen oxide emissions from the freight transport sector. The reason is that the national targets for nitrogen oxides and carbon dioxide for 2010 may not be reached with present policy. An agreement by participants in the project group was signed in May 1995, based on which an analysis of the problem, its causes, possible solutions and the role of industry and other institutions will be examined. The possibility for a covenant will be studied. In February 1995 the Environmental Performance Review was published. Concerning the transport sector it was recommended that the use of economic instruments (such as road pricing and fuel taxation) should be developed. The expansion of public involvement and the application of environmental impact assessment in the definition of transport policies were also recommended.

(h) *Slovakia*: The energy policy and strategy up to the year 2005 was approved by the Government in 1993. It defines the national carbon dioxide emission reduction target of 20 per cent based on 1988 and also aims at reducing emissions of other substances in line with international commitments.

(i) *Sweden*: In January 1995 a parliamentary committee was commissioned to propose a plan for the development of the national communication network with the aim of reaching a sustainable transport system. The committee is to report in late 1996.

(j) *Turkey*: A means of reducing air pollution in large cities and urban areas has been the introduction of alternative sources of energy, in particular natural gas. Natural gas is utilized in power plants and industries and its introduction into the heating sector is planned. Existing gas systems in the two largest cities have been converted to natural gas and a new network is under development to alleviate the severe air pollution episodes in the winter. Gas will also be made available in three other cities. Also, the increased use of new and renewable energy sources is being studied. At present the most important is solar energy, which is widely used for water-heating in coastal zones. The use of geothermal energy for district heating has also helped to solve some local air pollution problems.

(k) *European Community*: Recent activities in the energy sector include the ratification of the Energy Charter and the adoption of regulations on the characteristics of biodiesel. In the field of transport, the European Auto-Oil Programme has been conducted to identify a cost-effective package of measures to reduce mobile source emissions. The programme takes into consideration future air quality objectives and evaluates the emission reduction potential of technical measures and traffic management.

D. Administrative structures: national and local authorities

In order to be effective, national air pollution abatement policies must be implemented by administrative institutions. The institutional structures and mechanisms for enforcement and support in each country depend on national systems and traditions of public administration, including the division of competences between central, regional and local authorities. Most Parties to the Convention ensure central policy formulation and coordination for air pollution abatement at the national level by a combination of executive and operational bodies with consultative or advisory ones. The following subparagraphs summarize recent changes in the administrative structures or new information reported since the Major Review.

(a) *Denmark*: After parliamentary elections in 1994, the ministries of environment and of energy were merged into a single Ministry of Environment and Energy. This Ministry now also includes the Danish Energy Agency, the Mineral Resources Administration for Greenland and the Danish Forest and Landscape Research Institute. The main task of the Energy Agency is to ensure that energy production, supply and consumption develop in a social, safe and environmentally responsible manner. The Agency also maintains relations with national and international partners in the field of energy.

(b) *France*: The bulk of the regulatory and policing tasks in connection with regulations governing atmospheric discharges and their monitoring and surveillance are the responsibility of the regional directorates for industry, research and the environment, which report to the prefect in each region.

(c) *Turkey*: The Ministry of Environment has sole responsibility to coordinate and prepare regulations and to cooperate with other ministries and agencies as required. On air pollution control this includes close cooperation with the Ministry of Energy and Natural Resources, the Ministry of Health, and local authorities. The Ministry of Health, which has local branches in all provinces, is responsible for air quality measurements. At the local level municipal authorities also carry responsibility for atmospheric protection.

III. NATIONAL POLICY MEASURES

Policy measures have been grouped into four categories: regulatory measures, economic instruments, measures related to technology, and the monitoring and assessment of effects. As such a categorization may be difficult in some cases, cross-references have been indicated where appropriate. In practice a mix of instruments will prevail and the different types of measures should be considered complementary. For instance, whereas regulations tend to allow a more direct control of polluting sources and hence reduce the uncertainty of the policy result, economic instruments generally bring about emission reductions in a more cost-effective manner.

A. Regulatory provisions

1. Ambient air quality standards

Ambient air quality standards or target levels frequently serve as a reference base for other standards (emissions, fuel quality, control technology) designed to achieve a given desirable level of air quality. Ambient air quality is usually defined in terms of maximum yearly, daily and hourly average concentrations of specified pollutants, sometimes also as maximum concentrations for shorter-term episodes. The legally binding force of standards may be differentiated, for instance, as between stringent health-related standards and indicative nuisance standards. Standards related to health effects for sulphur dioxide, nitrogen dioxide, ozone, particulates and lead were listed in table 11.1 of the 1994 Major Review. Standards related to ecological effects for sulphur dioxide and ozone were given in table 11.2. Besides these, several Parties have established standards for other pollutants considered potentially harmful to health and the environment. The following subparagraphs update the information contained in the 1994 Major Review.

(a) *Czech Republic*: The standards for air quality outside urban areas are given by the set of ambient air limits in the Measure to the Clean Air Act. In addition, regulations have been prepared in connection with this Act through a decree of the Ministry of the Environment to regulate air pollution sources during smog hazard conditions.

(b) *France*: The norms applied in France stem directly from the European directives. These norms are due for modification over the coming five years, in the light of the forthcoming adoption of the Framework Directive on air quality by the European Union. In the summer of 1994 the country introduced a procedure for informing and warning the population in the Paris region concerning not only ozone levels but also levels of sulphur dioxide and nitrogen dioxide. The thresholds are as follows:

—180 μg/m³ and 360 μg/m³ for ozone (hourly average)
—300 μg/m³ and 400 μg/m³ for nitrogen dioxide (hourly average)
—350 μg/m³ and 600 μg/m³ for sulphur dioxide (hourly average).

These procedures will be extended by stages to the other French cities. In addition to the appeal to motorists not to use their vehicles during the alert periods, this procedure undoubtedly promoted greater awareness of air pollution problems among the population.

(c) *Germany*: As ground-level ozone frequently exceeded normal levels (in particular the health and vegetation protection thresholds of the relevant EC directive) over large areas during the summer months in recent years, a smog alarm regulation providing for temporary measures has been prepared. This measure against peak ozone levels has been taken while recognizing that only substantial (70-80 per cent) long-term reductions in the

emissions of the ozone precursors (NO$_x$ and VOCs) can safeguard against ozone episodes even under unfavourable meteorological conditions. Large-scale experiments have shown that locally isolated emission reductions do not suffice. The main element of the smog alarm regulation is a ban on the operation of high-emission motor vehicles over large areas. It is to be imposed when ozone concentrations reach a level of 240 μg/m³ (1 hour mean) and are expected to remain at this level for the following day. Low-polluting vehicles (e.g. those equipped with a closed-loop three-way catalytic converter) as well as vehicles used for special purposes (such as public transport) will be exempted from the ban.

(d) *Liechtenstein*: The ambient air quality standards for sulphur dioxide are 100 μg/m³ for the short (1/2 h) and medium (24 h) term and 30 as an arithmetic yearly average. For nitrogen dioxide they are 100 μg/m³ for the short term, 80 for the medium term and 30 for the yearly average. For ozone there are standards of 120 μg/m³ for the short term (1 h) and 100 for the medium term (98 per cent of 1/2 h averages during a month).

(e) *Luxembourg*: Following recent changes in legislation, air quality standards for nitrogen dioxide (NO$_2$), sulphur dioxide and suspended particulates, as well as thresholds for air pollution by ozone, are as follows:

Quality standards for NO$_2$ (in μg/Nm³)

Limit value in ambient air	200 (98th percentile of hourly average values recorded throughout the year)
Guide values in ambient air	50 (50th percentile of hourly average values recorded throughout the year)
	135 (98th percentile of hourly average values recorded throughout the year)

Limit values for SO$_2$ and suspended particulates associated with SO$_2$ (in μg/Nm³)

Period	SO$_2$	Suspended particulates Value associated with SO$_2$	
Year	80	>40	Median of daily average values during the year
	120	≤40	
Year	250	>150	98th percentile of daily average values during the year
	350	≤150	
Winter (1 October-31 March)	130	>60	Median of daily average values during the winter
	180	≤60	

Limit values for suspended particulates in μg/Nm³ (not associated with SO$_2$)

Period	Limit value
Year	80 (median of daily average values during the year)
Year	130 (98th percentile of daily average values during the year)
Winter (1 October-31 March)	250 (median of daily average values during the winter)

Thresholds for ozone concentration in the air	
Threshold for protection of health	$110 \, \mu g/m^3$ over 8 hours
protection of vegetation	$200 \, \mu g/m^3$ over 1 hour
	$65 \, \mu g/m^3$ over 24 hours
provision of information to the population	$180 \, \mu g/Nm^3$ over 1 hour
Threshold for alerting the population	$360 \, \mu g$ per m^3 over 1 hour

(f) *Russian Federation*: As of January 1995 threshold values or safe exposure guideline levels have been established for more than 1,900 substances and compounds, including isomeric polychlorinated dioxins and furans.

(g) *Turkey*: The ambient air quality standards for sulphur dioxide are $900 \, \mu g/m^3$ for the short (hourly) and 400 for the medium (daily) and 150 for the long term (yearly). For nitrogen dioxide they are $300 \, \mu g/m^3$ for the short term, and 100 for the yearly average. For ozone a standard of $240 \, \mu g/m^3$ for the short term has been set. For particulates there are medium-term standards of $300 \, \mu g/m^3$ and long-term values of 150. Finally, the long-term standard for lead is $2 \, \mu g/m^3$.

(h) *Ukraine*: Environmental safety regulations contain maximum permissible concentrations (MPCs) of pollutants in the air, maximum permissible emissions (MPEs) to the atmosphere, and environmental technical regulations, which directly limit emissions from various types of process equipment. MPCs and approximate safe levels of exposure have been set for 1,330 substances. Work has now been completed on establishing MPEs for air pollutants. MPEs have been set for virtually all enterprises.

(i) *European Community*: The new Council Framework Directive on Ambient Air Quality Assessment and Management, which is expected to be adopted in 1995, is intended to provide a more coherent, consistent and effective means for achieving air quality standards and objectives. It will replace the existing directives on sulphur dioxide, nitrogen oxides, lead and black smoke. The Framework Directive aims to define the principles of a strategy to: establish objectives for air quality; assess air quality in a uniform manner; provide air quality information to the public; and maintain good and improve poor air quality. It covers: the types of objectives (including guidelines for selecting pollutants); assessment methods; actions/measures required if the objectives are not met (short-term plans, including time schedule and public information); and information requirements. The European Commission will propose air quality objectives for: sulphur dioxide, nitrogen dioxide/oxides, black smoke, suspended particulate matter, and lead before 1997; for ozone in accordance with the directive 92/72 and for carbon monoxide, cadmium, acid deposition, benzene, PAH/BaP, arsenic, fluoride and nickel before the year 2000.

2. Target loads or deposition standards

Target loads play a similar role to that of ambient air quality standards in determining the basis for other policy measures. They are often established on the basis of scientifically determined critical loads (see sect. D.2).

(a) *Belgium*: The deposition target (guide value) for cadmium is currently $20 \, \mu g/m^2$ per day.

(b) *Canada*: The objective for wet deposition of sulphur is now 20 kg/ha per year. This target load was established to protect moderately sensitive aquatic ecosystems. The objective is at present under revision and a move towards critical loads, taking into account the effects of both nitrate and sulphate deposition on sensitive aquatic and terrestrial ecosystems, is being considered.

(c) *Luxembourg*: There is at present no legislation stipulating limit values for deposition. However, the environment administration applies the German TA Luft standards where possible.

(d) *Slovenia*: In 1994, in implementing ordinances of the Environmental Protection Act, deposition limits have been set as a daily average for: dust $350 \, \mu g/m^3$; lead $100 \, \mu g/m^3$; zinc $400 \, \mu g/m^3$; and cadmium $2 \, \mu g/m^3$.

3. Fuel quality standards

The regulation of sulphur content in fuels is a major element in emission control policies in the ECE region. At present, most Contracting Parties regulate the sulphur content of fossil fuels. The maximum permissible content is generally specified separately for heavy, medium, light and extra-light fuel oil, as well as for gas oil, coke and coal. In a few countries, fuel quality standards are uniformly applied nationwide. Another approach is for them to be more stringent in large urban areas and specially sensitive zones than in rural areas. In addition, in some cases, special attention is given to standards for the lead content of petrol. National standards for the different fuel types were summarized in table 12, fuel quality standards, of the 1994 Major Review. Some new information is provided below.

(a) *France*: The norms applied in France stem from stipulations laid down at the Community level. At the same time, French regulations allow for the imposition of a requirement to use fuels with a lower sulphur content—very low-sulphur (1 per cent) or very very low-sulphur (0.5 per cent) heavy fuel oil—as a temporary or permanent measure in certain polluted areas.

(b) *Germany*: As of July 1994 the ordinance on small firing installations has been amended to also regulate the sulphur content of lignite for use in small firing installations in the form of briquettes and originating from the former GDR. The same requirements as for hard-coal briquettes have become applicable. Through the addition of an additive, such as lime, the sulphur contained in the fuel has to be bound in the combustion residues to such an extent that the sulphur dioxide remaining in the waste gas is equivalent to that obtained when fuel with a maximum permissible sulphur content of 1 per cent by weight is used.

(c) *Greece*: The sulphur content in diesel fuel is currently 0.2 per cent.

(d) *Hungary*: The maximum permissible aromatic content of petrol has been reduced to 3 per cent in volume.

(e) *Ireland*: The national standard for benzene in petrol (both leaded and unleaded) is 5 per cent. The Department of the Environment is currently discussing with the oil companies how the benzene content of petrol could be further cut on a voluntary basis. In addition, regulations give effect to the EU Directive 93/12/EEC relating to the sulphur content of certain liquid fuels. They prohibit the marketing of gas oil with a sulphur content of more than 0.2 per cent (by weight) effective 1 October 1994 and, as of 1 October 1996, that of diesel fuel with a sulphur content of more than 0.05 per cent (by weight) for use in motor vehicles. The EU Stage 1 Directive on emissions of volatile organic compounds from petrol distribution is being transposed into Irish law.

(f) *Liechtenstein*: The sulphur content in light fuel oil is limited to 0.2 per cent, the use of medium and heavy fuel oil in combustion has been prohibited. The maximum sulphur content for hard coal is 1 per cent and for lignite 3 per cent. Regular leaded petrol has been banned.

(g) *Luxembourg*: The most recent modification of the legislation on fuel quality was contained in a regulation of 1 December 1993 restricting the sulphur content of diesel oils to 0.2 per cent (by mass).

(h) *Netherlands*: Legislation to implement EC directive 93/12/EEC on the use of low-sulphur gas oil entered into force in October 1994. The sulphur limit for gas oil is now 0.2 mg/m^3 and will be reduced to 0.05 mg/m^3 as of October 1996.

(i) *Norway*: New regulations on the sulphur content in fuels will enter into force in 1995. In the cities of Oslo and Drammen the use of residual oil will be forbidden. Corresponding to the EC directive 93/12/EEC, the maximum sulphur content in gas oil will be 0.2 per cent.

(j) *Slovenia*: In October 1995, the diesel sulphur content standard for road vehicles will be lowered to 0.2 per cent. In the future, all fuels will have to meet quality requirements corresponding to the standards set by the European Community in 1993, limiting the sulphur content of diesel fuel to 0.05 per cent by the year 2000.

(k) *Sweden*: Since March 1995 leaded fuel has been banned. In 1994 leaded fuels were phased out from the market in accordance with a voluntary agreement between the Swedish Petroleum Industries Association and the Swedish Environmental Protection Agency.

(l) *Turkey*: The sulphur content in light fuel oil is limited to 0.9 per cent, in medium fuel oil to 2 per cent and in heavy fuel oil to 2.5 per cent. The maximum sulphur content for hard coal is 0.6 per cent and for lignite 2.5 per cent.

(m) *Ukraine*: The sulphur content of fuel oil is regulated by a Ukrainian State standard. The maximum permissible sulphur content is 2 per cent for light oil; 3.5 per cent for medium oil and over 3.5 per cent for heavy oil. For industrial centres with a high level of pollution, use is made of fuel oil with a sulphur content of up to 2 per cent. Maximum permissible emissions for mobile sources and vehicles are being brought into effect through the setting of State and sectoral standards. The use of unleaded petrol is expanding. Emissions from mobile sources have been reduced due to the wide use of diesel engines, the use of liquefied and compressed natural gas by road transport, the monitoring of the toxicity of emissions and the adjustment of the fuel systems.

(n) *European Community*: The European Commission will submit a proposal to the Council on the reduction of the sulphur content of fuel derived from petroleum, such as heavy oils, bunker fuel, and aviation kerosene, to supplement directive 93/12/EEC. It will also present a proposal on a fuel quality directive.

4. Emission standards

Standards for the control of air pollutants either set maximum permissible quantities, for specific sources and for specified pollutants such as sulphur dioxide and nitrogen oxides, or require specific technological controls to be applied. Emission standards can be set industry by industry or plant by plant or on the basis of national emission standards for specific pollutants. Quantitative emission standards may be expressed in various forms: e.g. mass of pollutant per unit volume of flue gas; parts of pollutant per million or billion parts of flue gas; mass of pollutant emitted per unit energy output or fuel input. Table 13, Emission standards, of the 1994 Major Review summarizes national standards for SO_2, NO_x and VOCs for the different source categories. The following subparagraphs give an update.

(a) *Austria*: In an amendment to the Ordinance on Steam Boilers, stricter emission limits for nitrogen oxides, carbon monoxide, particulates and ammonia have been set and off-heat boilers have been included as a new source type. The emission standards for nitrogen oxides (in g NO_2/Nm^3) for new plants (power generation and steam boilers) as of January 1996 will be:

Capacity (MW$_{th}$)	Solid fuels	Fuel oil			Wood	
		Extra light	Light	Medium & heavy	Waste wood	Natural wood
< 10	0.40	0.15	0.40	0.45	0.50	0.30 or 0.25
10-50	0.35	0.15	0.35	0.35	0.35	0.20
> 50	0.20	0.15	0.10	0.10	0.20	0.20

For gaseous fuels the limits will be 0.125 for small (< 3 MW$_{th}$) and 0.10 g/Nm^3 for larger installations. For carbon monoxide the following standards have been set: for solid fuels, 1.0 for small (< 1 MW$_{th}$) and 0.15 g/Nm^3 for larger plants; for liquid fuels 0.10 for small (< 1 MW$_{th}$) and 0.08 g/Nm^3 for larger plants; for liquefied gas 0.10 g/Nm^3; for natural gas 0.08 g/Nm^3; and for wood 0.25 for small (0.1-5 MW$_{th}$) or 0.10 g/Nm^3 for larger plants. For particulates the emission limits are: for solid fuels 0.15 for small (< 2 MW$_{th}$) and 0.05 g/Nm^3 for large plants; for heavy and medium fuel oil 0.060 for small (< 30 MW$_{th}$), 0.050 for medium (30-50 MW$_{th}$) and

0.35 g/Nm^3 for larger plants; for light fuel oil 0.050 for small (< 30 MW$_{th}$) and 0.35 g/Nm^3 for larger plants; for extra light fuel oil 0.03 g/Nm^3; for gas 0.005 g/Nm^3; and for wood 0.25 for small (0.1-5 MW$_{th}$) and 0.10 g/Nm^3 for larger plants. Finally for steam boiler plants using ammonia/ammonium as a deNOx facility limits have been set at 0.03 g/Nm^3 for small (< 50 MW$_{th}$) and 0.01 g/Nm^3 for larger installations.

(b) *Belgium*: The emission limit values for 1994 for the regions of Flanders and Wallonia are given in the tables below.

BELGIUM—WALLONIA

Sulphur dioxide

Category of sources	Description	Emission limits	Other information
Power generation Unit: g/Nm^3	Small plants 50≤P<100 MW$_{th}$	2 to 1.7 (1.3 on 1/1/1998)	Solid fuel
		1.7 (1.3 on 1/1/1998)	Liquid fuel
		0.035 to 2.0	Gaseous fuel
	Medium-sized plants 100≤P<300 MW$_{th}$	1.2 to 1.7 (1.3 on 1/1/1998	Solid fuel
		1.7 (1.3 on 1/1/1998)	Liquid fuel
		0.035 to 2.0 (1.3 on 1/1/1998)	Gaseous fuel
	Large plants P≥300 MW$_{th}$	0.250	Solid fuel
		0.250	Liquid fuel
		0.035 to 2.0 (1.3 on 1/1/1998)	Gaseous fuel
Industrial operations Unit: g/Nm^3	Foundries	0.5	
	Refineries	0.5	
	Steelworks	0.5	
	Cement works	0.5	
	Paper pulp plants	0.5	
	Gasworks	0.5	

Volatile organic compounds

Type of sources	Emission limits	Other information
Stationary sources Unit: g/Nm^3	0.150 (overall standard)	(Expressed in terms of TC)
Mobile sources Unit: g/km	EC directive	
Fuel volatility Unit: g/test, per cent RVP		
Evaporation from organic solvents		

Nitrogen oxide

Category of sources	Description	Emission limits	Other information
Power generation Unit: g/Nm^3	Small plants $50 \leq P < 100$ MW$_{th}$	0.400 to 0.950 (0.650 on 1/1/1998)	Solid fuel
		0.200 to 0.575 (0.450 on 1/1/1998)	Liquid fuel
		0.100 to 0.425 (0.350 on 1/1/1998)	Gaseous fuel
	Medium-sized plants $100 \leq P < 300$ MW$_{th}$	0.200 to 0.950 (0.650 on 1/1/1998)	Solid fuel
		0.150 to 0.575 (0.450 on 1/1/1998)	Liquid fuel
		0.100 to 0.425 (0.350 on 1/1/1998)	Gaseous fuel
	Large plants $P \geq 300$ MW$_{th}$	0.200 to 0.950 (0.650 on 1/1/1998)	Solid fuel
		0.150 to 0.575 (0.450 on 1/1/1998)	Liquid fuel
		0.100 to 0.425 (0.350 on 1/1/1998)	Gaseous fuel
Industrial operations Unit: g/Nm^3	Nitric acid plants	0.450	
	Fertilizer plants	0.500	
	Petrol-fuelled light-duty vehicles	European standards	
	Petrol-fuelled heavy-duty vehicles	European standards	
	Diesel-fuelled light-duty vehicles	European standards	
	Diesel-fuelled heavy-duty vehicles	European standards	

BELGIUM—FLANDERS

Sulphur dioxide

Category of sources	Description	Emission limits	Other information
Power generation Unit: g/Nm^3	Small plants $50 \leq P < 100$ MW$_{th}$	2.0 1.7	Solid fuel Liquid fuel
	Medium-sized plants $100 \leq P < 300$ MW$_{th}$	1.2 1.7	Solid fuel Liquid fuel
	Large plants $P \geq 300$ MW$_{th}$	0.4 0.4	Solid fuel Liquid fuel
Industrial operations Unit: g/Nm^3	Foundries	0.8	
	Refineries	2.0 (< 1998), 1.3 (> 1998)	
	Steelworks	0.5	
	Cement works	0.5	General emission limit value
	Paper pulp plants	0.5	
	Gasworks	0.5	

Volatile organic compounds

Type of sources	Emission limits	Other information
Stationary sources Unit: g/Nm^3		
Mobile sources Unit: g/km	European standards	
Fuel volatility Unit: g/test, per cent RVP		
Evaporation from organic solvents		

Nitrogen oxide

Category of sources	Description	Emission limits	Other information
Power generation Unit: g/Nm^3	Small plants $50 \leq P < 100$ MW$_{th}$	0.650 0.45	Solid fuel Liquid fuel
	Medium-sized plants $100 \leq P < 300$ MW$_{th}$	0.650 0.450	Solid fuel Liquid fuel
	Large plants $P \geq 300$ MW$_{th}$	0.650 0.450	Solid fuel Liquid fuel
Industrial operations Unit: g/Nm^3	Nitric acid plants Fertilizer plants	0.50 0.50	General emission limit value
	Petrol-fuelled light-duty vehicles	European standards	
	Petrol-fuelled heavy-duty vehicles	European standards	
	Diesel-fuelled light-duty vehicles	European standards	
	Diesel-fuelled heavy-duty vehicles	European standards	

(c) *Bulgaria*: The emission standards for electricity and heat generation plants are being revised in line with the obligations under the 1994 Oslo Protocol and EC regulations. Standards for nitrogen oxides and hydrocarbons are being developed.

(d) *Canada*: Emissions from major stationary sources are mostly regulated by site- or sector-specific permits or limits issued by provincial environmental agencies. National minimum performance limits have been established for new stationary sources such as steam electric power plants, combustion turbines, and large reciprocating engines, and are being developed for industrial boilers and furnaces, cement and lime kilns. Light-duty vehicles are built according to national standards which, by agreement with vehicle manufacturers, follow new vehicle standards being phased in in the United States beginning with the model year 1994. Similarly, by agreement with leading manufacturers of heavy-duty diesel engines, engines sold in Canada have to meet the same emission standards as those built for the United States market and will track new standards as these are phased in.

(e) *Croatia*: Until the relevant national legislation is passed, standards adopted in European Community directives are applied.

(f) *France*: This is an area well-covered by French regulations, which include numerous provisions that are more stringent than Community rules. Community directives on the following subjects have been taken up in French regulations:

—Large combustion plants;

—Vehicles;

—Incineration of household wastes and dangerous wastes (currently being transposed).

In addition, limit values exist for all industrial plants (order of 1 March 1993), as well as for certain specific installations (orders relating to glassworks and cement works). The corresponding norms for new installations are as follows:

	Sulphur oxides	Nitrogen oxides	VOC	Others
Combustion -> 50 MW -< 50 MW	Directive on large combustion plants 3 400 mg/m^3 (in the process of adoption)	Directive on large combustion plants (in the process of adoption)	-	
General rule	300 mg/m^3	500 mg/m^3	150 mg/m^3 (20 mg/m^3 for carcinogens)	Cd+Hg+Tl 0.2 mg/m^3 As+Se+Te 1 mg/m^3 Other metals 5 mg/m^3 NH$_3$ 50 mg/m^3

	Sulphur oxides	Nitrogen oxides	VOC	Others
Glassworks	Equivalent electricity, gas or very low-sulphur fuel oil	500 to 1 500 mg/m^3 depending on the type of oven Target: 500 mg/m^3	-	*All metals* 5 mg/m^3 *except for special types of glass* Sb:3 mg/m^3 As^g:5 mg/m^3 As^p:1 mg/m^3 Cr, Pb, Va: 5 mg/m^3 Co, Ni, Se, Cr^v: 1 mg/m^3 *NH$_3$* General rule
Cement works	500 mg/m^3 other than exceptions	1 200 to 1 800 mg/m^3 depending on type of process used	-	*Heavy metals* General rule

An instrument in the process of adoption will regulate emissions from new combustion plants with capacities of between 2 and 20 MW. An equivalent instrument, also in preparation during 1995, is due to cover turbines and engines. For existing plants, the regulations vary with the technical and economic characteristics of each plant, with the aim of ensuring the greatest possible compliance with the limit values applicable to new plants. It should be noted in particular that 11 thermal power plant units are to be equipped with systems for downstream desulphurization, upstream desulphurization and upstream denitrification. All plants already use fuels with a sulphur content of under 1 per cent. The 13 oil refineries are also scheduled to reduce their SO$_2$ emissions by 50 per cent by the year 2000. All these regulatory provisions (the general order, the order on glassworks, the order on cement works and the circular on thermal power plants) were adopted in 1993.

(g) *Ireland*: All relevant EU directives, including Directives 93/59/EC and 94/12/EC on controls of emissions of air pollutants from motor vehicles, have been implemented with effect on the due dates. Manufacturers have the option of complying with either until 1 January 1997, when Directive 94/12 becomes mandatory. As a result of implementing these directives, emissions of nitrogen oxides, carbon monoxide and hydrocarbons will be significantly reduced and all new petrol engined vehicles can only be refuelled with unleaded petrol. The 1987 Air Pollution Act and the 1993 Municipal Waste Incineration regulations give effect to the EU Directive 89/369/EEC on the prevention of air pollution from new municipal waste incineration plants. They prescribe limit values for specified pollutants and give directions as to the best practicable means of limiting the emissions concerned. The regulations also specify conditions in relation to emission limit values and the monitoring and operating of plants.

(h) *Liechtenstein*: Emission limits have been set for about 150 substances and for 40 types of installations of industry and small and medium-sized firms. The limits for nitrogen oxides for light-duty vehicles (petrol or diesel) is 0.62 g/km, for diesel-fuelled heavy-duty vehicles it is 9 g/kWh. VOC emission limits for stationary sources range from 20 to 150 mg/m^3 depending on the substances. For mobile sources a VOC limit of 0.25 g/km is applied.

(i) *Luxembourg*: There are no general emission standards which limit the total emissions of specific pollutants on a national scale. However, a law and a ministerial circular exist which enable the State authorities to prescribe emission standards for point sources, i.e. individual facilities. These are the law of 9 May 1990 on dangerous, insalubrious or inconvenient facilities and the ministerial circular of 27 May 1994 providing for the use of the best available technology through the setting of recommended thresholds for emissions into the air from industrial plants and cottage industries.

(j) *Turkey*: Emission standards have been set for stationary sources. For sulphur dioxide (in g/Nm^3) in the power generation sector they are: 2.0 (solid fuel), 1.7 (liquid fuel) and 1.0 (gas) for small and medium-sized installations (< 300 MW$_{th}$ or < 100 MW$_{th}$ for gas) and 1.0 (solid fuel), 0.8 (liquid fuel) and 0.06 (gas) for larger plants. For cement plants a limit of 0.4 has been set. For nitrogen oxides an emission limit of 0.8 g/Nm^3 (solid and liquid fuels) and 0.5 (gas) has been set for plants larger than 50 MW$_{th}$ (solid and liquid fuels) and 100 MW$_{th}$ (gas).

(k) *Ukraine*: Emission standards have been set for all substances for each stationary source. The criterion in setting the maximum permissible emissions for each particular source has been that, bearing in mind their dispersion and transformation in the atmosphere and the likely future development of industry, pollutant emissions from that source and all the other sources in the town or settlement concerned should not result in surface concentrations that will breach air quality standards. In numerical terms, the ambient air quality standard is that quantity of a pollutant in the air that will not cause pathological changes in humans, animals or vegetation. Maximum permissible concentrations have been established for most air pollutants. Sectoral emission standards are being set for some industries.

(l) *European Community*: The new directive 94/67/EEC adopted in 1994 on the incineration of hazardous waste sets emission standard monitoring requirements and rules of compliance for the incineration of waste in industrial plants. An emission limit value of 0.1 nanogram dioxin per m^3 will come into force no later than January 1997. This standard will help to meet the target of the Fifth Action Programme to reduce emissions of dioxin by 90 per cent by the year 2005. For mercury and other heavy metals stringent emission limit values are set. Special standards are fixed for waste incinerating plants which are not especially designed for that purpose, such as cement kilns. With directive 04/12/EEC the Council amended the basic legislation on emissions of motor vehicles (70/220/EEC). The new directive presents a new set of standards for light-duty vehicles as stringent as those for passenger cars and enters into force in 1996/97. Additionally, this directive requires the European Commission to come forward with a proposal for new emission standards for passenger cars effective from 2000 onwards.

5. Licensing of potentially polluting activities

A common regulatory procedure in ECE member countries is for the government to authorize (by certification or licensing) the initial operation of potential sources of air pollution and to impose specific environmental requirements (such as quantitative emission standards or a specific technology for pollution abatement) for the continued operation of such sources. Licences may also prescribe a time schedule for progressively more stringent requirements, thus forcing industries to develop more advanced technology. In the case of mobile sources, certification may additionally specify regular in-service controls and inspections—including pollution control tests—as a prerequisite for continued operation (e.g. of vehicles). With regard to stationary sources, most ECE countries require operating permits either for all major sources of pollution or for specified categories of activities or establishments. New information reported since the 1994 Major Review is summarized below.

(a) *Austria*: A treaty between the *Länder* (provinces) has been concluded to set minimal requirements and emission limits for the licensing of small-scale firing installations.

(b) *Belgium*: The Flemish environmental legislation, called VLAREM, follows the same principles as the proposed EU Directive on integrated pollution prevention and contains two parts. VLAREM 1 introduces the obligation to deliver an integrated environmental licence to all installations before they are allowed to start activities. All installations are classified into three categories according to their environmental impact. Environmental licences are issued by the provincial authorities for installations with the highest impact (class 1) and by municipal authorities for installations with a smaller impact (class 2). For the smaller installations (class 3) an announcement to the municipal authorities is sufficient. VLAREM 2 prescribes objectives for noise, water quality, air quality, soil and groundwater protection and emissions standards for all industrial installations. The emission standards reflect the most recent assessment for best available techniques. For air emissions, the German TA Luft standards are introduced in VLAREM 2.

(c) *Canada*: Large stationary emission sources such as industrial boilers and refineries are licensed by provincial or regional authorities. Most large projects are licensed following a detailed environmental assessment and these permits are periodically reviewed and, as needed, made more stringent. New light-duty vehicles must comply with federal emission standards. In-use inspection and maintenance programmes are being considered in densely populated jurisdictions and a programme is in place in the Vancouver region of British Columbia.

(d) *France*: Many fixed installations are subject to an authorization procedure. This applies in particular to all combustion plants with a capacity of over 20 MW, all incineration, steel-making and refining plants, cement works and glassworks with a capacity of over 1,500 tons a year. It should be noted that these regulations already apply to many agricultural activities—agro-food industries and stock-raising of a certain size—or other types of non-industrial polluting activities, notably the dumping of wastes. All these plants are subject to individual orders which are consistent with national and international rules. These orders may specify conditions for monitoring and surveillance of discharges as well as their impact on the environment. Similarly, for vehicles, the French procedure is in keeping with the procedure set out in Community rules. However, under a ministerial order adopted in 1994, the frequency of technical inspections of all vehicles will be increased starting in 1996, instead of 1998, as stipulated in the latest directive to be adopted, and remedial work will be required on polluting vehicles.

(e) *Ireland*: The Environmental Protection Agency established in 1993 has a wide environmental brief, including the licensing of a broad range of activities. Licences will be based on an Integrated Pollution Control system taking account of impacts on air, water and soil, and of noise.

(f) *Norway*: All enterprises having furnaces with a capacity above 500 kg fuel oil or emissions exceeding 10 kg SO_2 per hour must apply for a permit. Plants emitting 6 to 10 kg SO_2 per hour are obliged to notify the authorities. The EC regulations concerning waste and hazardous waste incineration are being implemented. The Norwegian Pollution Control Authority is preparing national emission guidelines for smaller plants (below 50 MW_{th}), burning gas oil and solid fuels. The guidelines will be completed during the autumn of 1995 and cover emissions of particles, sulphur dioxide, nitrogen oxides, carbon monoxide and other pollutants. They will deal with residue handling, plant operation and control measures. These guidelines will be used by the authority when licensing potentially polluting activities.

(g) *Turkey*: Based on the Air Quality Control Regulation, facilities which have significant negative effects on human health and the environment have to obtain licences for their operation. The permits are granted in accordance with the provisions of the Public Health Law taking into account either the views of the Ministry of Environment or those of the local environmental board depending on the type of installation.

(h) *Ukraine*: Under current Ukrainian law, every enterprise has to obtain a permit for the emission of air pollutants. The permits are issued on the basis of the maximum permissible emissions (MPEs) to the atmosphere, with account being taken of the implementation of measures to bring emissions down to the levels set in those standards. In addition, a health-protection zone is established for each enterprise, in accordance with health regulations. In selecting building sites, designs are examined from the point of view of the potential for long-range transboundary air pollution.

6. Product-oriented requirements and labelling

A product-oriented approach is followed in particular to control VOC emissions, for instance in the case of solvents. In such cases product reformulation will be required in order to replace polluting substances by less harmful substitutes. Alternatively, the labelling of potentially polluting products may be required.

(a) *Canada*: The Environmental Choice Programme, established by Environment Canada, permits the use of an "EcoLogo" under licence on products and services that meet stringent environmental criteria. The criteria are contained in guidelines that are developed in consultation with industry, environmental groups, universities and independent technical and scientific advisers.

(b) *Czech Republic*: In the framework of the programme of the Ministry of Environment for awarding the Eco-Logo "Environmentally Friendly Product", the Minister of the Environment has approved a Guideline containing the conditions for awarding the Eco-Logo to hot-water gas-fuelled boilers with atmospheric burners and with forced-air burners with an output of up to 200 kW, for hot-water flow-through gas-fuelled boilers with an output of up to 50 kW and for solid-fuel hot-water boilers and local furnaces with an output of up to 200 kW. The conditions for awarding the Eco-Logo require high combustion efficiency and low pollutant emissions (NO_x, CO, solid substances and hydrocarbons). In addition, guidelines have been approved for awarding the Eco-Logo to water-dispersible coatings, which must not contain aromatic and chlorinated hydrocarbons or halogenated solvents.

(c) *Ireland*: The country participates in and is fully committed to the EU eco-labelling scheme. The Irish competent body has been designated. If and when criteria for further product groups are established, Ireland will be in a position to consider applications for the eco-label.

(d) *Norway*: Regulations requiring all harmful substances and products to carry warning labels have been in force since 1982. These regulations aim at preventing injury to human health. The labelling also covers solvents and indirectly contributes to a reduction in VOC emissions. Labelling of environmentally hazardous substances in accordance with EC regulations (directives 67/548/EEC and 88/379/EEC) will enter into force in 1995. These labelling requirements will then focus on health and environmental effects.

(e) *Slovenia*: A proposal for an ordinance based on the EC regulations on chemical substances is under preparation. For the import of products like CFCs, hal-

ons, and carbon tetrachloride an import permit is already required and products have to be properly labelled.

(f) *European Community*: In 1992, an EC eco-label award scheme was agreed upon for products that do not seriously damage the environment during the entire cycle from the raw material to the waste elimination stages. Criteria for eco-labels for paints and varnishes as well as for hairsprays are currently being developed.

7. Other regulatory measures

(a) *Netherlands*: The Interim Act on Ammonia and Livestock entered into force in early August 1994, replacing the Guidelines on Ammonia and Livestock. To prevent an increase in ammonia emissions near sensitive objects, the Act sets rules to assess the ammonia deposition originating from livestock farms. Local authorities can grant permits to farms that do not have sufficient permits under the Environmental Protection Act. The period for the application of the Interim Act is limited to five years. On the other hand, the Act gives local authorities the possibility to pursue their own policy by developing ammonia reduction plans, which aim at emission and deposition reductions while enlarging the possibilities for farm development. The Note on Manure and Ammonia Policy (Third Phase) described in the 1994 Major Review, which sets the Government strategy up to the year 2000, has been reviewed and amended by the Integral Note on Manure and Ammonia, which was issued to Parliament in October 1995.

(b) *Ireland*: For some years urban smoke levels in Cork have threatened to exceed national and EU air quality standards. The Air Pollution Act and Regulations (1994) prohibit the marketing, sale and distribution of bituminous coal in Cork City and adjacent areas. They were introduced as a precautionary measure with effect from 13 February 1995. Enforcement of the ban is the responsibility of the local authorities concerned. The controls are similar to those in place in Dublin since 1990.

B. Economic instruments

Economic instruments refer to any measures that aim to reduce the pollution burden through financial incentives and hence will lead to a transfer of resources from the owner of a polluting source to the community or will directly change relative prices.

1. Emission charges and taxes

Emission charges or taxes require payment in relation to the amount of a given pollutant or the characteristics of the pollutant. In some countries a system of fines is used if standards are exceeded or licence requirements are not met. If rates are high enough to internalize the full social costs of a polluting activity at its source, they are in line with the polluter-pays principle: they compel the polluter to absorb those costs as part of his production costs. The rates charged should therefore be proportionate to the estimated cost of the environmental damage caused by the emissions. Charges or taxes are often linked to energy consumption, as emissions can in many cases be directly related to energy use; in those cases where they are applied to the price of products, such as fuels, they have been included under subsection 2. The following subparagraphs summarize recent developments and new information.

(a) *Belgium*: In the framework of the legislation on economic expansion an "ecology criterion" was established to grant environmentally friendly investments in Flanders financial support of 15 per cent over and above the basic expansion support. Three categories of investments are considered, namely investments which have a positive effect on savings of raw material, on energy saving and on care for the environment. The investments are compared with a list of technologies which might be considered. The list helps the objective application of the ecology criterion but does not prevent other ecological projects, i.e. those not listed, from being considered for the support. In 1994, a total of BF 9.7 billion was recognized as ecological investments, 3 billion of which were earmarked for energy-saving.

(b) *Czech Republic*: In a 1994 amendment to the Clean Air Act charges were extended to cover also small pollution sources. The rates (Czech koruna per year) are the following:

Fuel/capacity	0-50 kW	50-100 kW	100-200 kW
Fuel oil (0.3-1 per cent S content)	400-800	800-1250	1250-1700
Light fuel oil	700-1400	1400-2100	2100-2800
Other fuel oils	1000-2000	2000-3000	3000-4000
Hard coal	500-1000	1000-1500	1500-2000
Graded brown coal	1050-2100	2100-3100	3100-4200
Brown coal for the power industry	2000-4000	4000-6000	6000-8000
Coal sludges	10000	10000-20000	20000-40000

Coke, wood, gas, kerosene, fuel oil (< 0.3 per cent S) are exempted from the charge.

(c) *Denmark*: A tax of DKr 100 per ton of CO_2 was introduced in 1992. In connection with the tax reform of 1993/94 it was decided to put greater emphasis on "green" taxes. The Energy Package was adopted by the Parliament in June 1995 and is to be evaluated in 1998. The CO_2 tax now distinguishes three areas of energy consumption in industry: (1) heavy processes, (2) light processes and (3) room heating. For category (1) the tax of DKr 5 per ton of CO_2 in 1996 is gradually increased to DKr 25 in 2000, while reimbursement is possible through an agreement on an energy-saving action plan implying a financial burden of DKr 3 per ton of CO_2. For category (2) the tax of DKr 50 per ton in 1996 will be raised to DKr 90 in 2000, and firms with a high energy intensity can be partially reimbursed if they agree to an energy-saving action plan. For category (3) taxes will be raised up to 600 DKr per ton of CO_2, which will also be applicable to private households. A tax of DKr 10 per kilogram of sulphur was introduced. The tax will be gradually phased in from 1996 to the year 2000 through a scaling-down of the basic allowance charged on different fuel types. The tax is reimbursed as sulphur is filtered away or bound in other products. On electricity generation the tax will be levied only in the year 2000. Until then the electricity tax is being gradually increased.

(d) *France*: France instituted a special tax on air pollution as early as 1985. This tax, introduced for a five-year period, has been regularly extended and broadened, in 1990 and again in 1995. Originally, only emissions of sulphur compounds were taxed, at a rate of 130 francs per ton emitted. In 1990, the coverage was extended to oxycompounds of nitrogen and hydrochloric acid, and the rate was raised to 150 francs per ton emitted. The system was extended once again in 1995; this time, it was broadened to cover volatile organic compounds, and the rate was raised to 180 francs per ton. It applies to all combustion plants with a capacity of over 20 MW, incineration plants with a capacity of over 3 tons of wastes per hour, and any plant emitting more than 150 tons of one of the taxable pollutants over a year. The revenue from this tax is used essentially for three purposes:

—Assistance to industries in installing facilities for the treatment or reduction of emissions of the taxable pollutants;

—Assistance to industries developing technologies for the reduction of emissions of the taxable pollutants;

—Funding the national air quality surveillance system.

(e) *Liechtenstein*: The introduction of a tax on organic solvents, and a combined CO_2/energy tax is planned.

(f) *Poland*: In December 1994 emission charge rates were substantially raised to keep pace with the rate of inflation.

(g) *Russian Federation*: By Government decision of August 1992, a procedure was introduced for determining the charges, *inter alia*, for stationary sources. The following basic rates, which are periodically adjusted to allow for inflation (in 1994 they were to be increased by a coefficient of 10), were laid down: sulphur dioxide, 330 roubles per ton/year; nitrogen dioxide, 415 roubles per ton/year, and ammonia, 415 roubles per ton/year. If the emission exceeds the standard rates, while remaining within the permitted limits, the payment is increased fivefold. For emissions exceeding the permitted limits, the amount is again increased fivefold. The payment rates can be differentiated in the light of the environmental situation in particular areas.

(h) *Sweden*: The charge on emissions of nitrogen oxides from large combustion plants has recently been modified to include all plants producing more than 25 GWh (before 50 GWh). The charge is SKr 40 per kg NO_x emitted and is refunded in proportion to the total amount of energy produced.

(i) *Ukraine*: The first step in the application of economic instruments to resource use was the introduction in January 1992 of environmental pollution charges. Basic standard charges have been fixed for emissions from stationary sources. Basic standard charges for emissions from mobile sources depend on the quantity and type of fuel in question (highest price per ton for leaded petrol, medium price for diesel, lowest price for unleaded petrol). Air pollution charges are levied for the entire volume of emissions, including emissions below the maximum permissible emissions (MPEs). Restraining measures include the issue of emission permits without which an enterprise is not entitled to emit harmful substances into the air. In addition, enterprises are checked systematically by the Ministry of Environmental Safety. Thus, 200 enterprises have been identified as principal ambient air polluters and are checked at least once a year. More than 16,000 enterprises were checked in 1994. The inspections revealed that many of them were exceeding MPEs. Fines can be levied for failure to comply with air protection measures.

(j) *European Community*: The modified proposal for a Council Directive introducing a tax on carbon dioxide and energy introduces elements of flexibility which, from January 1996 to January 2000, will allow EC member States to apply this tax within a harmonized structure, on the basis of common parameters. During this period of transition they will preserve a certain freedom for determining the rates applied in this framework.

2. Product charges, taxes, and tax differentiation, including fuel taxes

Product charges and taxes are often a substitute for emission charges and taxes, where the level of pollution resulting from a particular activity can better be quantified by the amount of a product that is used in, or results from, the activity. Thus these charges or taxes are applied to the price of the product that causes pollution as it is manufactured, used as a factor of production or consumed. In general a classification system will be required distinguishing products according to the pollution that they may cause. Tax differentiation is especially used in relation to fuel taxes in order to set an incentive for the consumption of low-pollution fuels. Differentiation of fuel and car taxes have, for instance, been widely applied to encourage the introduction of unleaded petrol. Some recent changes in fuel taxes and new information provided by Parties are presented below.

(a) *Belgium*: The legislation on income tax grants enterprises in Flanders which carry out environment or

energy-saving investments a supplementary tax deduction of 10 per cent. Together with the basic deduction of 4.5 per cent for energy-saving investments or investments in research into environmentally friendly activities, the extra deduction amounted to 14.5 per cent.

(b) *Bulgaria*: In 1993 an additional import tax was introduced on old vehicles. The income goes to the National Environmental Fund. In 1994 the excise tax for unleaded petrol was lowered. In 1995 Parliament introduced a tax of 10 per cent on petrol and diesel fuels. Three per cent of the income has been earmarked for ecological projects in the mountainous regions through a special fund of the Ministry of Environment.

(c) *France*: The principal product charges with a potential impact on air pollution are excise duties imposed on fuels. Excise duties on high-octane petrol and unleaded high-octane petrol in France are among the highest in Europe. It should also be noted that excise duties on high-octane petrol were raised by around 50 centimes (10 per cent) two years ago. Duty on diesel fuel was also raised, by about 40 centimes (almost 20 per cent).

(d) *Hungary*: A number of tax or import duty reductions have been set to encourage air pollution control in the transport sector. The weight tax for cars equipped with catalytic converters has been reduced by 50 per cent. Engines to replace two-stroke engines can be imported free of duty and consumption tax. Catalytic converters and some monitoring devices are free of import duty. Alternative fuels, such as natural gas, are subject to lower consumption tax rates.

(e) *Liechtenstein*: Fuel taxes were increased by Sw F 0.20 per litre in 1994. The taxes on motor vehicles were raised in accordance with the respective weight. The introduction of a tax on fuel oil with a sulphur content of more than 0.1 per cent is planned.

(f) *Netherlands*: According to the second National Environmental Policy Plan, fuel excise taxes will as of 1996 at a minimum be kept stable in real terms, i.e. nominally raised at least in line with the rate of inflation. To attain the CO_2 target a proposal for a regulatory tax on energy for small consumers planned for 1996 has been proposed.

(g) *Norway*: The tax differential between leaded and unleaded petrol has been modified and, since January 1995, depends on the amount of lead in petrol. For a lead content of up to 0.05 g/litre the tax on petrol is Nkr 0.22 per litre, for a larger lead content it is Nkr 0.67 per litre. The carbon tax on oil, levied on the amount and type of fuel consumed, has been changed as of January 1995. The rates (in Nkr) are:

Fuel	Rate	Equivalent to
Petrol	0.83 per litre	358 per t CO_2
Autodiesel	0.415 per litre	156 per t CO_2
Residual oil	0.415 per litre	134 per t CO_2
Distilled fuel oil	0.83 per m^3	156 per t CO_2
Natural gas (on the continental shelf)	0.83 per litre	355 per t CO_2
Diesel (on the continental shelf)	0.83 per litre	312 per t CO_2
Condensate (on the continental shelf)	0.83 per litre	480 per t CO_2
Coal and coke in combustion	0.415 per kg	171 (coal)/130 (coke) per t CO_2

The consumption of mineral oil in the manufacture of paper and paper products and metal industry has a lower rate of Nkr 0.2075 per litre. The consumption of coal and coke in cement and metal production is not taxed.

(h) *Russian Federation*: By Government decision of August 1992, charges on emissions from mobile sources have been levied in the form of a fuel tax. The amounts to be paid in roubles per ton (or per 1,000 m^3 of gas), allowing for the inflation coefficient, are as follows: ethylated petrol AI-93, 646, A-76, A-72, 425; non-ethylated petrol AI-93, 170, A-76, A-72, 187; diesel fuel, 357; compressed natural gas, 153; liquefied gas, 197. In the absence of data on the amount of fuel consumed, the amounts to be paid depend on the type of transport (in thousands of roubles per vehicle): motor car, 45.9; lorry and bus with petrol engine, 68; gas-powered motor vehicle, 23.8; lorry and bus with diesel engine, 42.5; construction and road machines and agricultural equipment, 8.5; passenger diesel locomotive, 275.4; goods diesel locomotive, 363.8; shunting diesel locomotive, 42.5; passenger vessel, 255; cargo vessel, 340; auxiliary fleet vessel, 102. If catalytic converters or filters are used the amounts payable are reduced by the following coefficients: for motor vehicles using non-ethylated petrol and gas fuel, 0.05; and for other kinds of transport, 0.10.

(i) *Sweden*: In November 1994 a tax incentive was introduced to promote a better quality of unleaded petrol. The petrol has a lower content of sulphur and benzene and a lower vapour pressure.

(j) *Turkey*: To encourage the use of unleaded petrol its price was cut in April 1993.

(k) *European Community*: The Transport White Paper emphasizes the need to internalize increasingly and more appropriately external costs (including environmental costs). Council Directive 93/89/EEC on annual tax and the charging of transport infrastructure costs to lorries, marks a first step in this direction. The review of EC legislation on excise duties for mineral oil (directive 92/82/EEC), due in 1994, takes into account environmental considerations.

3. User and administrative charges

User and administrative charges constitute payments for specific services supplied by public authorities. User charges are, for instance, applied to cover the cost of waste collection or sewerage treatment and hence are less frequently found in the field of air pollution. Administrative charges are usually applied to cover the cost of licensing and monitoring activities by authorities. Some new information is provided below.

(a) *France*: The Government imposes two charges of this type on designated installations: the single charge, imposed to cover the costs of the authorization procedure for such installations; and the designated installations charge, which corresponds to the discharges and nuisance caused by the worst-polluting installations. However, the latter generates lower revenues than the special tax on air pollution and helps to finance monitoring of such installations.

4. Emission trading

Emission trading requires a definition and distribution of property rights to environmental media such as the air, or parts thereof ("bubble concept"). Instead of determining the emission limits for a specific plant, a sum of emissions for a specific area or source category is defined and transferable rights to emit (permits) are distributed among enterprises. These can then decide whether to reduce emissions and sell permits or whether to buy additional permits. Consequently, emission reductions will occur where they are the cheapest to accomplish. No new information concerning emission trading has been reported since the 1994 Major Review.

5. Subsidies and other forms of financial assistance

Subsidies are another form of economic instrument although they usually do not conform with the polluter-pays principle. Such measures include low-interest loans, accelerated write-off allowances, cash grants for investments in pollution abatement equipment, research and development. Those subsidies are granted to individuals or enterprises either indirectly (in the case of tax rebates, which result in a loss of tax revenue) or as disbursements from the budget of environmental departments, sometimes through special funds set up for the purpose. The reduction or removal of subsidies that induce polluting activities above the optimal level is also important in some countries.

Financial assistance may not be in violation of the polluter-pays principle if it is used to speed up a period of transition and is thus limited in time, or if it consists of payments for positive externalities. For instance, general energy-saving measures, which incidentally also lead to a decrease in the emissions of sulphur and nitrogen compounds, receive government support in a number of countries.

(a) *Bulgaria*: The National Environmental Protection Fund receives financing from charges, fines, taxes, donations and other means. Also 5 per cent of the receipts from privatization have been directed towards the Fund since 1994. In 1994 projects amounting to 182 million leva were financed through the Fund in the form of grants for the municipalities and interest-free credits for private firms. The expected receipts of the Fund for 1995 are 532 million leva. Indirect State subsidies are provided in the form of preferential duties for devices for ecomonitoring and emission reduction and waste treatment installations and equipment.

(b) *France*: For many years a system of tax deductions has been in force for operations that can lead to energy savings in all sectors, either through deductions from income tax, or through a system of companies whose funds are derived from the savings achieved.

(c) *Greece*: The government subsidies for the installation of pollution control filters on vehicles were abolished.

(d) *Norway*: In 1995 subsidies for energy efficiency will focus on information, education and grants for the introduction of energy efficiency technology. At the local level, the Government is establishing regional energy efficiency centres. By the end of 1995 some ten centres will be established. They will offer advice on energy efficiency measures and information on the use of different energy carriers and their tariffs. For 1995 Government has granted Nkr 35 million for information and training and Nkr 20 million for demonstration projects.

(e) *Sweden*: A temporary investment subsidy to encourage home-owners to install accumulators connected to small-scale wood furnaces has been introduced. The subsidy covers 30 per cent of total investment.

(f) *Turkey*: An Environmental Pollution Prevention Fund was formed under the Environmental Law. Up to 45 per cent of the expenditure necessary for the prevention of environmental pollution and the improvement of the environment may be provided by credits with a maximum duration of 20 years from the Fund. The Fund is administered by the Ministry of Environment and may be used for the following purposes: research, clean-up, education, training, technology development, purchase of equipment, reforestation and studies of ecosystems.

(g) *European Community*: In December 1993, the European Commission adopted a set of new environmental aid guidelines on the basis of which it will judge State aid for environmental protection. The new guidelines will cut back on aid given for adapting existing plants, but as subsidies must form part of the second best solution, they allow for a higher rate of aid for investment that goes significantly beyond current environmental protection requirements.

C. Measures related to emission control technology

1. Technology requirements in legislation and regulations

A common approach for ensuring that appropriate control technology is applied to different polluting activities, is to require the use of the so-called "best available technique" (BAT), "state-of-the-art technology" or "best practicable means". In some countries, the concept of best available or practicable technology is explicitly stated in environmental legislation, whereas others stipulate the use of best practicable technology in the permits and licences required in order to undertake potentially polluting activities. Several countries introduce best practicable technology into their air pollution policies by setting emission standards on the basis of the technical performance of those technologies. The following subparagraphs summarize new developments.

(a) *Belgium*: In the past, measures to reduce the emissions often focused on problem areas where high ambient concentrations occurred. During the last few years, more stringent emission limit values were imposed on the major industrial installations in the three regions. This policy has been generalized by the publication in Flanders of the new environmental legislation VLAREM II, described above, and, in Wallonia, of a decree which imposes BATNEEC on all polluting industries. Wallonia is preparing a technical instruction on air quality based on the principles set in TA-luft. In this way an important step forward has been made towards improving air quality and reducing emissions.

(b) *Canada*: In general, emission limits are set as performance standards and increasingly more flexible approaches are being explored. Even though performance standards do not impose a specific technology, they are derived with the knowledge that a given level of performance is achievable using one or more proven and feasible techniques.

(c) *Czech Republic*: The concept of best available techniques not entailing excessive cost (BATNEEC) is laid down in the Clean Air Act and forms the basis for setting emission limits in the relevant regulations. It is also applied in the procedure for issuing construction and operation permits for potentially polluting emission sources.

(d) *Hungary*: Legislation to be issued in 1996 will change the basis for emission limit values from the system of regional emission limit values to technological emission limit values. The requirements for important technological processes and emission control methods have been drawn up on the basis of BAT taking into account the specific conditions of the country.

(e) *Norway*: Emission standards are not generally applied, but specific limit values are set case by case based on BAT as defined either at the national or the international level. In addition, local conditions are taken into account.

(f) *Turkey*: The Ministry of Environment regularly prepares circulars containing detailed provisions and emission control methods. These are sent out once a year to the local authorities, which have to follow the provisions and technology requirements.

(g) *European Community*: The Directive on Integrated Pollution Prevention and Control proposed by the European Commission in 1993 and likely to be adopted in 1995, aims at integrating industrial licensing and ensuring that best available techniques are used to control and prevent air pollution to all affected media as well as to reduce waste and consumption of energy and raw material. Permits must set emission limit values based on what is achieved through the use of BAT. The Directive sets out a general definition and criteria for the determination of BAT, whereas the actual determination is at the EC member States' discretion. Member States are obliged to inform the European Commission of the emission limit values imposed and the BATs from which these have been derived.

2. Control technology requirements for stationary sources

Technological requirements form an integral part of the Sulphur, NO_x and VOC Protocols. The following paragraphs summarize recent changes concerning the requirements on emission control technologies for major stationary sources of air pollution.

(a) *Czech Republic*: Control technology requirements for stationary sources are based on the emission standards for new sources set by the Measure to the Clean Air Act. For solid fuels special emission limits are set for SO_2, NO_X, CO, hydrocarbons, and other substances. Emission limits for existing sources are set individually by the relevant air protection authority; however, by the end of 1998 all sources must conform to the emission limits for new sources.

(b) *Slovakia*: New and modernized sources must meet current emission standards based on BATNEEC. Existing sources (i.e. those authorized before October 1991) have until December 1998 to comply with limits set for new sources.

(c) *Slovenia*: According to ordinances to the Environmental Protection Act implemented in 1994 emission standards for new sources apply immediately, while existing sources have to meet standards in a period defined through the rehabilitation programme.

(d) *European Community*: The Council adopted directive 94/63/EEC on VOC emissions from the petrol distribution chain (stage I) that applies to storage, loading and transport of petrol between terminals and between terminals and service stations. It lays down provisions for the design and operation of storage, unloading equipment, mobile containers and storage equipment at service stations. A proposal is under preparation on the control of refuelling losses (stage II) which will require the setting in place of vapour recovery equipment at service stations with a high throughput of petrol, or in sensitive areas designated by EC member States.

3. Control technology requirements for mobile sources

As regards motor vehicles, most ECE countries use as a basis for national approval procedures the 1958 ECE Regulations annexed to the Agreement Concerning the Adoption of Uniform Conditions of Approval and Reciprocal Recognition of Approval for Motor Vehicle Equipment and Parts currently applied by 27 ECE member States. The development of these Regulations meets a continuing need to improve road safety and to reduce the damage to the environment at a time of continuing growth in motor vehicle traffic. There are currently 97 Regulations. They are amended, or supplemented, in response to the concerns of society and to the changing technology. Under the uniform system so established, approval of vehicle types and components (including the conformity of production tests related to pollutant emissions for instance) is a prerequisite for marketing, including the import of vehicles or parts manufactured outside the ECE region, and in some countries it also suffices for the operation of imported vehicles. Some countries additionally require regular in-service controls and inspections as a prerequisite for continued operation of vehicles in road traffic. As of 18 September 1995, Regulations concerning the emission of pollutants by motor vehicle engines were in force in the following countries:

(*a*) Regulation No. 15 (with four series of amendments and a supplement, gradually being replaced by Regulation No. 83) on gaseous pollutants from passenger cars and light-duty vehicles equipped with a positive-ignition engine using leaded petrol (or with a compression-ignition engine): Croatia; Romania; Russian Federation; and Yugoslavia;

(*b*) Regulation No. 24 (with three series of amendments) on visible pollutants from compression-ignition engines or from vehicles equipped with such engines: Belarus; Belgium; Croatia; Czech Republic; Finland; France; Germany; Hungary; Italy; Luxembourg; Netherlands; Poland; Romania; Russian Federation; Slovakia; Slovenia; Spain; United Kingdom; and Yugoslavia;

(*c*) Regulation No. 40 (with one series of amendments) on gaseous pollutants from motor cycles equipped with a positive-ignition engine: Belarus; Belgium; Croatia; Czech Republic; Finland; France; Germany; Hungary; Italy; Luxembourg; Netherlands; Norway; Poland; Romania; Russian Federation; Slovakia; Slovenia; United Kingdom; and Yugoslavia;

(*d*) Regulation No. 47 (in its original version) on gaseous pollutants from mopeds equipped with a positive-ignition engine: Belgium; Croatia; Czech Republic; Finland; France; Germany; Hungary; Italy; Luxembourg; Netherlands; Norway; Poland; Romania; Russian Federation; Slovakia; Slovenia; United Kingdom; and Yugoslavia;

(*e*) Regulation No. 49 (with two series of amendments) on emissions of pollutants (gaseous and particulate) by compression-ignition engines for heavy-duty vehicles: Belarus; Belgium; Croatia; Czech Republic; Finland; France; Germany; Hungary; Italy; Luxembourg; Netherlands; Poland; Romania; Russian Federation; Slovakia; Slovenia; United Kingdom; and Yugoslavia;

(*f*) Regulation No. 67 (in its original version) on equipment for using liquefied petroleum gas in motor vehicle propulsion systems (safety requirements): Belarus; Belgium; Czech Republic; Finland; Hungary; Italy; Netherlands; Norway; Poland; Romania; Slovakia; and United Kingdom;

(*g*) Regulation No. 83 (with two series of amendments) on the approval of vehicles with regard to the emission of gaseous and particulate pollutants by the engine according to the engine fuel requirements: Belarus; Belgium; Czech Republic; France; Germany; Hungary; Italy; Luxembourg; Netherlands; Poland; Romania; Slovakia; Slovenia; Spain; United Kingdom; and Yugoslavia;

(*h*) Regulation No. 84 (in its original version) on the approval of power-driven vehicles equipped with internal combustion engines with regard to the measurements of fuel consumption: Austria; Belgium; Czech Republic; Finland; France; Germany; Hungary; Italy; Luxembourg; Netherlands; Norway; Poland; Romania; Slovakia; Slovenia; Spain; United Kingdom; and Yugoslavia;

(*i*) Regulation No. 85 (in its original version) on the approval of power-driven vehicles equipped with internal combustion engines with regard to the measurement of the net power: Belgium; Czech Republic; Finland; France; Germany; Hungary; Italy; Luxembourg; Netherlands; Norway; Poland; Romania; Slovakia; Slovenia; Spain; United Kingdom; and Yugoslavia.

(*j*) Regulation No. 96 (in its original version) on the emissions of pollutants (gaseous and particulate) by the compression-ignition engines to be installed in agricultural and forestry tractors: entry into force for Italy and the United Kingdom on 15 December 1995;

(*k*) The additional draft regulation on the emissions of carbon dioxide and the fuel consumption of passenger cars was adopted in 1994 and will be enacted by the Governments of France and Germany.

Requirements for mobile source emissions are also included in the Sulphur, NO$_x$ and VOC Protocols. The following paragraphs summarize the recent changes or new information since the 1994 Major Review on requirements set by Parties on emission control technologies for major categories of mobile emission sources.

(a) *Bulgaria*: A comprehensive long-term programme will be launched in order to introduce standards and emission control requirements for new and existing mobile sources which are in line with the EC requirements.

(b) *Greece*: Since 1995, an inspection system for the control of emissions from vehicles has been applied. All vehicles must bear a card indicating that the vehicle is in good working order and that it meets the emission limit values set for exhaust fumes.

(c) *Hungary*: New legislation in preparation will set requirements so that catalytic converters will have to be fitted to all new vehicles.

(d) *Netherlands*: On the basis of EC regulations type-approval, requirements for gaseous emissions have been set. Light motor vehicles (petrol and diesel) have to comply with the regulations of the directive 70/220/EEC, as amended by directive 92/12/EEC. Following this amendment a recent proposal by the European Commission will require more stringent limit values for light-duty commercial vehicles, implying that vehicles up to a weight of 1,200 kg will fall under the same requirements as passenger cars. The values for heavier light-duty vehicles will also become more stringent, but to a lesser extent, taking into account the technical specifications of these vehicles. Heavy motor vehicles (diesel) have to comply with regulations of directive 88/77/EEC amended by directive 91/542/EEC. Following technological developments, the latest proposal by the European Commission concerns the emissions of particulate matter from low-output diesel engines and would imply that the present maximum of 0.61 g/kWh would be reduced to 0.15 g/kWh as of October 1997. As of that date these engines would have to comply with requirements that apply to other diesel engines. In the second National Environmental Policy Plan, a scoping study for the installation of speed governors in private cars and vans has been proposed. Following this proposal, Parliament has requested a pilot project to be undertaken to experiment with speed governors in private cars.

(e) *Norway*: As of October 1993 the EC emission standard (directive 91/542/EEC A-level, "Euro 1") for heavy-duty vehicles became compulsory. The limit values are 4.9 g/kWh for carbon monoxide, 1.23 g/kWh for hydrocarbons, 9.0 g/kWh for nitrogen oxides and 0.4 g/kWh for particulates. From October 1996 (EC directive 91/542 B-level, "Euro 2") these standards will be 4.0 g/kWh for carbon, 1.1 g/kWh for hydrocarbons, 7 g/kWh for nitrogen oxides and 0.15 g/kWh for particulates. As of January 1995 the EC directives on vehicle emission standards for petrol-fuelled and diesel-fuelled passenger cars and for light-duty trucks came into force.

(f) *Turkey*: Emission limits reflecting the 1993 EC regulations for vehicles ("Euro 1 and 2") will gradually be introduced in the period between 1995 and the year 2000.

(g) *European Community*: Directive 93/59/EEC further tightened the emission regulations for light commercial vehicles with effect from 1993/94. Directive 92/6/EEC requires lorries (85 km/h) and buses and coaches (100 km/h) to be fitted with speed limiters. This also applied to a certain extent to existing vehicles. It is estimated that this measure will not only increase safety, but also reduce energy consumption (by at least 5 per cent). In the framework of existing EC legislation (77/143/EEC) on roadworthiness tests, directive 93/55/EEC introduced minimum standards for emission tests on all vehicles. Given that a large share of emissions is caused by a small number of badly maintained vehicles, it is expected that this directive will have a significant effect.

4. The availability of unleaded fuel

The 1994 Major Review concluded under this item that unleaded fuel was widely available. Most distribution facilities now offer unleaded fuel. In some countries leaded petrol or some type of petrol has been banned. In *Austria* only unleaded petrol has been available since November 1993; in *Canada* leaded petrol was phased out in the late 1970s-early 1980s and is no longer used in on-road vehicles; in *Germany, Liechtenstein* and *Switzerland* regular petrol is only available unleaded; in *Ireland* unleaded petrol is available in nearly 85 per cent of all outlets nationally; in *Sweden* leaded petrol has been completely phased out.

D. Monitoring and assessment of air pollution effects

1. Monitoring of air quality and environmental effects

Air pollution monitoring systems may deal with emissions, deposition, air and precipitation quality and the environmental effects related to air pollution. They may be designed to provide information on local problems or those on a regional scale. The following subparagraphs summarize recent changes or new information reported since the 1994 Major Review regarding national activities related to the systematic monitoring of the effects of air pollutants.

(a) *Austria*: A study of the Austrian Federal Environmental Agency deals with the atmospheric heavy metal depositions and the use of mosses as accumulation indicators. The results of the study will pave the way for the establishment of a nationwide monitoring programme in 1995 covering about 200 sites using the moss method. Another project of the Agency deals with the concentration of several organic and inorganic pollutants in precipitation at ten sampling sites. Besides the work for the International Cooperative Programme on Integrated Monitoring on Air Pollution Effects, numerous activities related to forest research and reported on in the 1994 Major Review are continuing.

(b) *Ireland*: Under the EU Directive 92/72/EEC a monitoring network for ozone has now been established. Monitoring is carried out by the Environmental Protection Agency at six locations around the country—Dublin, Cork, Galway, Monaghan, Wicklow and Kilkenny. There is also full access from the network to a further monitor at Logh Navar in Co. Fermanagh, Northern Ireland. Information on this pollutant and its precursors (NO_x and VOCs) is limited and the network is intended tò increase the information in this area, as a first step to specifying an air quality standard for this pollutant. The results of monitoring are transmitted automatically. The Directive stipulates a population information threshold of 180 $\mu g/m^3$ and a population warning threshold of 360 $\mu g/m^3$. In the event of an exceedance of either threshold, a public information system is activated as required by the Directive, whereby the population will be informed through weather bulletins on national television and radio, on local radio and weather forecasts in the national papers. The information provided to the public will include a forecast of the duration of the exceedance and the precautions to be taken by those at risk.

(c) *Liechtenstein*: Two stationary and one mobile monitoring stations are in operation covering all major air pollutants.

(d) *Turkey*: Monitoring activities have been extended to cover the whole country. At 73 locations and 76 provincial centres sulphur dioxide and suspended particles are being measured. In 1993 a new EMEP centre was established in Cubuk, near Ankara.

(e) *Ukraine*: The level of air pollution in Ukrainian towns is assessed from the analyses of air samples taken at 173 fixed points in 54 localities. The major part (60 per cent) of the information comprises data relating to such common harmful substances as dust, sulphur dioxide, nitrogen oxide and dioxide and carbon monoxide. The amounts of 38 pollutants, including heavy metals, in the air are measured. Air pollution is monitored in localities that are home to some 70 per cent of the country's population, including 44 cities of more than 100,000 inhabitants. The pollutants in atmospheric precipitation are identified at 33 points. The sulphate, chloride, hydrocarbonate, nitrate, ammoniated nitrogen, calcium, potassium and magnesium contents and the acidity of the precipitation are measured.

(f) *European Community*: In order to promote harmonization and the exchange of information on air pollutant monitoring in EC member States, the European Commission proposed a decision on establishing a reciprocal exchange of information and data from network and individual stations measuring ambient air pollution. The proposal recommends data for 34 pollutants (including SO_2, acidity, black smoke, PM_{10}, O_3, NO_2, NO_x, CO, H_2S, Pb, Hg, Cd, Ni, Cr, MN, As, CS_2, benzene, toluene, styrene, acrylonitrile, formaldehyde, trichloroethylene, tetrachloroethylene, dichloromethane, BaP, PAH, vinyl chloride, VOCs, PAN, and wet sulphur and nitrogen depositions) and averaging times. The data shall be made available to international bodies.

2. Research into air pollution effects and assessment of critical loads and levels

In addition to efforts within the framework of the Convention, there are numerous activities at the national level to assess the effects of air pollutants, in particular where these provide a basis for determining critical loads and levels within their territories. A summary of recent activities is given below.

(a) *Denmark*: A new report on the mapping of critical loads for sulphur and nitrogen, prepared by the Danish Environmental Research Institute, was published in October 1995. It describes the new methodology and its results and presents the exceedances of critical loads.

(b) *Liechtenstein*: Surveys are carried out sporadically on specific species (flora and fauna).

(c) *Netherlands*: After ten years of acidification research, the Priority Programme on Acidification ended in 1994. The results of the third and final phase became available in April 1995. The conclusions indicate a decrease in acid depositions in the Netherlands. Estimates of dry deposition were changed compared to earlier results and critical load values have also been updated. The effects of sulphur and nitrogen depositions, as well as the increasing threat through eutrophication to forest ecosystems, were highlighted.

(d) *Switzerland*: In 1992 work was started to apply more complex steady-state multi-layer soil models for assessing critical loads of acidity for forest soils. The results of this study were presented in the spring of 1995. The Swiss Federal Institute for Forest, Snow and Landscape and the Institute for Applied Plant Biology have used dynamic models to evaluate the critical loads of acidity. An in-depth evaluation of critical levels for ozone started in 1994 with an emphasis on vegetation. A comparison between measurements and calculations of the models will be conducted in a joint exercise with *Austria* and *Germany*.

IV. INTERNATIONAL ACTIVITIES

A. The Convention

Table 8 summarizes the status of the Convention on Long-range Transboundary Air Pollution and its related protocols.

(*a*) The Convention on Long-range Transboundary Air Pollution, adopted in Geneva on 13 November 1979 (E/ECE/1010), entered into force on 16 March 1983. As of 28 February 1995, with the accession to the Convention by the Republic of Moldova in June 1995, 39 States and the European Community were Parties to the Convention.

(*b*) The Protocol on Long-term Financing of the Co-operative Programme for Monitoring and Evaluation of the Long-range Transmission of Air Pollutants in Europe (EMEP), adopted in Geneva on 28 September 1984 (ECE/EB.AIR/11) and in force since 28 January 1988, has been ratified by 35 Parties.

(*c*) The Protocol on the Reduction of Sulphur Emissions or their Transboundary Fluxes by at least 30 per cent, adopted in Helsinki on 8 July 1985 (ECE/ EB.AIR/12) and in force since 2 September 1987, has been ratified by 21 Parties.

(*d*) The Protocol concerning the Control of Emissions of Nitrogen Oxides or their Transboundary Fluxes, adopted in Sofia on 31 October 1988 (ECE/EB.AIR/21) and in force since 14 February 1991, has been ratified by 25 Parties.

(*e*) The Protocol concerning the Control of Emissions of Volatile Organic Compounds or their Transboundary Fluxes was adopted in Geneva on 18 November 1991 (ECE/EB.AIR/30) and has been signed by 23 Parties and ratified by 13 Parties.

(*f*) The Protocol on Further Reduction of Sulphur Emissions was adopted in Oslo on 14 June 1994 (ECE/EB.AIR/40) and has been signed by 28 Parties and ratified by 3 Parties.

B. Activities aimed at improving the exchange of control technology

Besides activities carried out under the Working Group on Abatement Techniques, many Parties have taken measures to cooperate on air pollution control technology. They include the organization of training programmes, financial cooperation and trade measures aimed at improving the exchange of technology. The following subparagraphs present new activities in this field not reported on in the 1994 Major Review.

(*a*) *Canada* has held targeted workshops with international participation in pulp mill sludge combustion and ozone-depleting substance destruction technologies. Workshops were also held on biological gas cleaning and fine particles removal. Others may follow.

(*b*) The First North American Conference and Exhibition on Emerging Clean Air Technologies and Business Opportunities was successfully held in cooperation with the *United States* and *Mexican* Government Departments, in September 1994 in Toronto, *Canada*. A follow-up, "Clean Air 96", will be held in November 1996 in Orlando, *United States*. Environment *Canada* and the *United States* Environmental Protection Agency (EPA) meet annually to review research and development programmes in air pollution control.

(*c*) As part of the cooperation between *Finland* and the *Russian Federation*, the first desulphurization line was put into operation at the Kostamuksh ore-dressing combine in Karelia in 1994.

(*d*) A three-year project between *Hungary* and *Japan* was successfully finalized in April 1995. It covered the transfer of technology as well as air quality monitoring and modelling of heavily polluted regions.

(*e*) *Norway* (the Ministry of Environment and the Pollution Control Authority) has supported Cleaner Production Programmes in the *Czech Republic*, *Poland*, the north-western part of the *Russian Federation*, and in *Slovakia*. The main goals of the Programmes are to reduce pollution discharges and to achieve an optimal exploitation of natural resources and raw materials. In addition, *Norway* regularly conducts training programmes for advisers from central and eastern Europe.

(*f*) To reduce sulphur dioxide emissions, the reconstruction of the Pechenganikel combine (*Russian Federation*) has been planned. Based on tender, firms from *Norway* and *Sweden* will be offered contracts. The Government of *Norway* has provided a non-repayable grant of Nkr 300 million for the project.

(*g*) The *European Community* has, since 1992, conducted the LIFE programme. LIFE is a financial instrument which aims at contributing to the development and implementation of EC environmental policy. The European Commission has proposed the extension of LIFE to countries in central and eastern Europe, *inter alia*, to promote technologies and innovative approaches to environmental issues. Since 1992, the THERMIE programme has, through the Organization for the Promotion of Energy Technology (OPET), promoted energy-efficient technologies and provided expertise for the promotion of effective national energy infrastructures. The European Commission has proposed to change SYNERGY into a programme. The objective of the programme would be to improve the long-term world energy situation by providing support on energy policy development. Other programmes that conduct projects in countries in central and eastern Europe related to technology cooperation include the PHARE and TACIS programmes.

C. Other bilateral activities in the ECE region

Besides a number of multilateral initiatives reported on in the 1994 Major Review (chap. IV, sect. D), several Parties are engaged in bilateral programmes within the ECE region for air pollution abatement. Below some recent developments:

(*a*) *Austria* has concluded a memorandum of understanding on cooperation on environmental aspects of energy policy and the protection of the global climate with the *United States*.

(*b*) *Bulgaria* has concluded bilateral agreements on cooperation on the environment with *Denmark* and *Germany*. *Bulgaria* also has conducted negotiations with *Switzerland* on an agreement for a "debt-for-environment" swap. The agreement would cover about Sw F 20 million and set up a mechanism for managing these funds.

(*c*) In the framework of cooperation between *Denmark* and *Hungary*, a control strategy for emissions of VOCs in *Hungary* is being developed. In the framework of environmental cooperation between *Hungary* and *Norway*, a project on the development of a planning tool for air pollution control in the Pécs area has been concluded.

(*d*) Cooperation has started between the *Netherlands* and *Slovakia* based on a memorandum of understanding on air pollution.

(*e*) A joint project between *Luxembourg* and *Slovakia* on air pollution monitoring and public information in the city of Košice, *Slovakia,* has been started.

V. CONCLUSIONS

A. Status of implementation of the 1985 Helsinki Protocol on the Reduction of Sulphur Emissions or their Transboundary Fluxes by at least 30 per cent

Table 9 is based on the emission data in table 1 that had been submitted by Parties to the Convention by 31 August 1995. Parties to the 1985 Helsinki Protocol are shown shaded in table 9. Figure I shows sulphur emissions in 1993, or a previous year if no data were available for 1993, as a percentage of 1980 levels separately for Parties and non-Parties to the Protocol. One Party to the Protocol, Luxembourg, has not reported sulphur emission levels for the years after 1985. Two Parties to the Protocol, Belgium and Germany, have not yet reported emission data for 1993. One Party, the Russian Federation, reported only on sulphur emissions originating from stationary sources.

Taken as a whole, the 21 Parties to the 1985 Helsinki Protocol reduced 1980 sulphur emissions by 50 per cent by 1993 (using the latest available figure, where no data were available for 1993). In the whole of Europe, including non-Parties to the Protocol, that sum of emissions is well below 30,000 kt, which corresponds to a reduction of 47 per cent compared to 1980. Also individually, based on the latest available data, all Parties to the Sulphur Protocol have reached the reduction target. Also four non-Parties to the Protocol have achieved sulphur emission reductions of 30 per cent or more. Eleven Parties have achieved reductions of at least 60 per cent; two of these have actually reduced their sulphur emissions by 80 per cent or more.

Given that the target year for the Helsinki Protocol is 1993, it can be concluded that all Parties to that Protocol have reached the target of reducing emissions by at least 30 per cent.

B. Status of implementation of the 1988 Sofia Protocol concerning the Control of Emissions of Nitrogen Oxides or their Transboundary Fluxes

Table 10 is based on the emission data in table 2 that had been submitted by Parties to the Convention by 31 August 1995. Parties to the 1988 Sofia Protocol are shown shaded in table 10. Figure II shows nitrogen oxide emissions in 1994, or a previous year if no data were available for 1994, as a percentage of 1987 levels, or 1978 for the United States, separately for Parties and non-Parties to the Protocol. In some cases even data for emissions dating from before 1992 had to be used. Among the Parties to the Protocol, Luxembourg, Spain, and the European Community have not reported NO_x emission levels for the years after 1991; Luxembourg and the European Community have not reported emission data for 1987, their base year for the reduction target under the Sofia Protocol. For two Parties, Germany and Ireland, only 1992 data were available. The lack of data severely limits the possibility of evaluating the status of implementation.

Concerning the emissions of nitrogen oxides, the general reference year is 1987 with the exception of the United States, which chose to relate its emission target to 1978. For all Parties to the Convention overall emissions of NO_x had been stabilized by 1990 at the 1987 level and by 1993 (or an earlier year, where no figures are available for 1993) they had been reduced by 6 per cent. Taking the sum of emissions of Parties to the NO_x Protocol in 1993, or a previous year, where no recent data are available, also a slight reduction of 6 per cent compared to 1987 can be noted. Eighteen of the 25 Parties to the 1988 Sofia Protocol have reached the target and stabilized emissions at 1987, or in the case of the United States 1978, levels or reduced emissions below that level according to the latest emission data reported. Among the other cases two cannot be evaluated because of a lack of data for the base year and the five remaining Parties to the Protocol have increased emissions by 4 to 41 per cent above 1987 levels. Six Parties to the Convention (including one non-Party to the Sofia Protocol) have reduced NO_x emissions by more than 25 per cent. All but one of these are countries with economies in transition. It can also be noted that, in general, in southern Europe NO_x emissions have increased, in some cases significantly, above 1987 levels.

Concerning obligations under article 2, paragraph 2, of the Sofia Protocol, seven Parties (Belarus, France, Hungary, Luxembourg, Russian Federation, Ukraine, and United States) to the Protocol have not submitted any information on emission standards for NO_x. For other Parties the responses to the questionnaire under this item were incomplete. For those Parties that did submit data it can be concluded that emission standards have been set in most cases for new sources in the power generation sector. Fewer Parties have reported on emission standards for mobile sources. Concerning the availability of unleaded fuel, based on the responses received, it can be concluded that the requirement of article 4 has been largely fulfilled and unleaded fuel is available throughout the ECE region.

Based on available data, it can be concluded that eighteen Parties to the Protocol (Austria, Belarus, Bulgaria, Canada, Czech Republic, Denmark, Finland, France, Germany, Hungary, Liechtenstein, Netherlands, Norway, Russian Federation, Slovakia, Sweden, Switzerland, and Ukraine) have fulfilled their obligation un-

der article 2, paragraph 1, of the Protocol, to stabilize emissions at 1987 levels by 1994 at the latest. Another Party (United States) has specified the year 1978 as reference year and stabilized emissions at that level, but has, in addition, to limit its average emissions or transboundary fluxes for the period between 1 January 1987 and 1 January 1996 at levels of the year 1987. For the remaining six Parties no data for 1994 have yet become available and previous data do not show that they have fulfilled their emission reduction obligation. Concerning the obligation under article 2, paragraph 2, of the Protocol, no conclusions can at this stage be drawn from the available information.

Table 1. Emissions of sulphur (1980-2010*) in the ECE region
(thousands of tonnes of SO_2 per year)

	1980	1981	1982	1983	1984	1985	1986	1987	1988	1989	1990	1991	1992	1993	1994	2000*	2005*	2010*
Austria	397	·	·	242	·	195	·	152	122	93	90	84	76	71	74 a	78	·	·
Belarus	740	730	710	710	690	690	690	811	780	720	710	724	509	433	381	552	490	·
Belgium	828	712	694	560	500	400	377	367	354	325	317	324	304	294	·	248	232	215
Bosnia and Herzegovina	·	·	·	·	·	·	·	·	·	·	480	·	·	·	·	·	·	·
Bulgaria	2050	·	·	·	·	·	·	2420	2228	2180	2020	1667	1120	1422	1485	1374	1230	1127
Canada	4643	4291	3612	3625	3955	3692	3627	3762	3838	3695	3236	3245	3117	3008	2651	2833	2914	2914
Croatia	150	·	·	·	·	·	·	·	107	·	180	108	107	114	89	133	125	117
Cyprus	·	·	·	·	·	36	37	41	46	47	51	41	45	43	46	62	62	62
Czech Republic	2257	2341	2387	2338	2305	2277	2177	2164	2066	1998	1876	1776	1538	1419	1270	1128	902	632
Denmark	451	362	369	314	296	339	284	251	242	193	180	242	189	157	156	90	90	·
Finland	584	534	484	372	366	383	332	328	302	244	260	194	141	123	117	116	·	·
France	3338	2588	2490	2094	1866	1470	1342	1290	1226	1334	1298	1379 a	1238 a	1129 a	·	868	770	737
Germany	3166 b	3010 b	2843 b	2666 b	2578 b	2369 b	2230 b	1907 b	1218 b	945 b	5331	4176	3440	3156	2997	990	740	·
Germany, former GDR	4351	4433	4599	4682	5058	5365	5413	5444	5263	5254	·	·	·	·	·	·	·	·
Greece	400	·	·	·	·	500	·	·	·	·	510	·	·	·	·	595	580	570
Hungary	1633	1580	1545	1480	1440	1404	1362	1285	1218	1102	1010	913	827	756	741 a	898	816	653
Iceland	6	·	·	·	6	6	·	23 c	24 c	24 c	24 c	23 c	24 c	24 c	24 c	24 c	23 c	23 c
Ireland	222	192	158	142	142	140	162	174	152	162	178	179	161	157	155	155	·	·
Italy	3800 d	·	·	3150 d	2656 d	2244 d	2257 d	2274 d	2216 d	2001 d	1681	1643 a	1537 a	1490 a	·	1209	1042	·
Latvia	·	·	·	·	·	·	·	·	·	·	·	·	·	·	·	·	·	·
Liechtenstein	0.39	0.37	0.34	0.32	0.29	0.27	0.25	0.22	0.20	0.17	0.15	0.15	0.14	0.14	0.13	0.11	0.11	·
Lithuania	·	·	·	·	·	·	·	·	·	·	·	·	·	·	·	·	·	·
Luxembourg	24	·	·	14	·	17	·	·	·	·	14	·	·	15	12	4	·	·
Netherlands	490	464	404	323	299	261	264	263	250	204	205	195	170	161	147 a	92	·	56
Norway	141	127	110	103	95	97	91	74	67	59	54	45	37	36	35	34	·	·
Poland	4100	·	·	·	·	4300	4200	4200	4180	3910	3210	2995	2820	2725	2605	2583	2173	1397
Portugal	266 e	·	·	·	·	198	·	·	·	·	283	297	351	300	272	304	294	·
Republic of Moldova	·	·	·	·	·	·	·	·	·	·	·	·	·	·	·	·	·	·
Romania	·	·	·	·	·	·	·	1762	2397	1647	1504	1167	559	·	·	·	·	·

Table 1 (continued)

	1980	1981	1982	1983	1984	1985	1986	1987	1988	1989	1990	1991	1992	1993	1994	2000*	2005*	2010*
Russian Federation [f]	7161	6949	7090	6934	6503	6191	5707	5622	5145	4677	4460	4392	3839	3456	2983	4440	4297	4297
Slovakia	780	·	·	·	·	613	604	614	589	573	543	446	380	325	238	337	295	240
Slovenia	235	254	256	270	249	240	244	218	210	211	195	180	188	182	177	92	45	37
Spain	3319	·	·	2543	·	2190	1961	1903	1587	1950	2266	2223	2195	2071	·	2143	·	·
Sweden	508	431	371	305	296	266	272	228	224	160	136	112	103	101	97	100	·	·
Switzerland	116	·	·	·	84	76	·	·	·	·	43	41	38	34	31	30	30	30
Turkey	860	·	·	·	276	322	354	·	·	·	·	·	·	·	·	·	·	·
Ukraine	3849 g	3492	3427	3498	3470	3463	3393	3264	3211	3073	3782	2538	2376	2194	1715	2310	2310	2310
United Kingdom	4903 g	4441 g	4216 g	3864 g	3722 g	3727 g	3902 g	3897 g	3824 g	3724 g	3752 g	3563 g	3496 g	3184 g	2709	2320	1470	980
United States	23779	22512	21211	20618	21056	20833	20125	19877	20282	20388	20195	20194	19588	19856	19158 h	17382	15023	14093
Yugoslavia [i]	406	408	409	440	456	478	470	484	502	506	508	446	396	401	424	680	889	1135
European Community	25513 j	·	·	·	·	13626 j	·	·	·	·	·	·	·	·	·	8860 k	·	·

Source: data submitted by the Parties to the Convention on Long-Range Transboundary Air Pollution.

Emissions of sulphur: Notes

* Projections based on current reduction plans.
a Preliminary data.
b Figures apply to the Federal Republic of Germany as prior to 1989.
c Based on the IPCC-methodology.
d Preliminary data. No information as to whether nature is included (570 kt in 1990).
e Adopted from the 1994 Sulphur Protocol.
f Emissions from stationary sources only. Figures apply to the European part within EMEP.
g No information as to whether nature is included (9 kt in 1994).
h Adopted from the EPA report: EPA-454/R-95-011 (1995).
i Emissions from stationary sources only.
j CORINAIR total of 12 EC member States.
k 5th Community Programme of Policy and Action in Relation to the Environment and Sustainable Development.

Table 2. Emissions of nitrogen oxides (1980-2010*) in the ECE region
(thousands of tonnes of NO_2 per year)

	1980	1981	1982	1983	1984	1985	1986	1987	1988	1989	1990	1991	1992	1993	1994	2000*	2005*	2010*
Austria	246	.	.	241	.	245	.	234	226	221	222	216	201	182	177 a	155	.	.
Belarus	234	235	235	237	240	238	258	263	262	263	285	281	224	207	203	217	184	.
Belgium	442	.	.	.	.	315	307	321	335	347	343	347	350	350	.	.	.	.
Bosnia and Herzegovina	.	.	.	.	.	.	.	.	.	.	.	.	.	.	.	.	.	.
Bulgaria	.	.	.	.	.	.	.	416	415	411	376	273	260	238	327	380	350	290
Canada	1959	1907	1897	1884	1871	2038	2043	2131	2204	2188	2104	2003	1997	2006	2026	2061	2057	2085
Croatia	.	.	.	.	.	.	.	.	.	.	83	57	50	53	59	87	83	83
Cyprus	.	.	.	.	.	9	9	10	10	11	11	13	13	14	15	18	20	20
Czech Republic	937	819	818	830	844	831	826	816	858	920	742	725	698	574	369	398	.	.
Denmark	274	240	260	254	266	294	312	302	292	272	269	319	274	267	272	203	192	.
Finland	295	276	271	261	257	275	277	288	293	301	300	291	282	280	283	224	224	224
France	1823	1701	1688	1645	1632	1615	1618	1630	1615	1772	1585	1623 a	1597 a	1520 a	.	.	.	.
Germany	2926 b	2842 b	2817 b	2862 b	2923 b	2912 b	2943 b	2839 b	2745 b	2606	3071	2941	2913	2874 a	2872 a	.	2130	.
Germany, former GDR	731	726	692	685	706	738	740	759	748	760	.	.	.	.	.	.	.	.
Greece	.	.	.	.	.	306	.	.	.	.	.	.	.	.	.	.	.	.
Hungary	273	270	268	266	264	262	264	265	258	246	238	203	183	173	183 a	230	210	196
Iceland	13	.	.	.	12	12	.	18 c	19 c	19 c	20 c	21 c	22 c	23 c	22 c	21 c	20 c	19 c
Ireland	73	86	86	85	84	91	100	115	122	127	115	119	125	122	105	105	105	.
Italy	1480	.	.	.	1568	1741	1804	1904	1982	2035	2053	2067	2043	1997	.	2098	2060	.
Latvia	.	.	.	.	.	.	.	.	.	.	.	.	.	.	.	.	.	.
Liechtenstein	0.71	0.71	0.70	0.69	0.68	0.67	0.66	0.65	0.65	0.64	0.63	0.61	0.58	0.56	0.54	0.41	0.37	.
Lithuania	.	.	.	.	.	.	.	.	.	.	.	.	.	.	.	.	.	.
Luxembourg	23	.	.	21	.	22	.	.	.	.	23	.	.	25	21	19	.	.
Netherlands	583	575	562	555	573	576	587	599	602	584	575	575	566	543	526 a	249	.	120
Norway	184	177	183	188	204	215	229	237	229	232	230	221	220	229	225	161	.	.
Poland	1229	.	.	.	.	1500	1590	1530	1550	1480	1279	1205	1130	1120	1105	1345	.	.
Portugal	.	.	.	.	.	96 d	.	.	.	.	215	227	247	246	253	.	.	.
Republic of Moldova	.	.	.	.	.	.	.	.	.	.	.	.	.	.	.	.	.	.
Romania	.	.	.	.	.	.	.	369	253	1753	883	805	443	.	.	.	.	.

Table 2 (*continued*)

	1980	1981	1982	1983	1984	1985	1986	1987	1988	1989	1990	1991	1992	1993	1994	2000 *	2005 *	2010 *
Russian Federation e	1734	1915	2002	1976	1879	1903	1871	2653	2358	2553	2675	2571	2298	2269	1995	.	.	.
Slovakia	.	.	.	.	.	.	.	197	.	227	227	212	192	184	173	.	.	.
Slovenia	48	49	49	48	49	50	54	53	55	54	53	50	51	57	66	45	38	31
Spain	950	.	.	937	.	839	854	892	892	992	1178	1227	1251	1223	.	892	.	.
Sweden	454	417	412	401	411	426	432	437	432	418	411	410	402	398	392	312	303	311
Switzerland	170	.	.	.	177	179	.	.	.	.	166	160	153	145	140	117	110	113
Turkey	0.3	0.3	0.3	0.3	0.3	0.4	0.4	0.4	0.4	0.5	0.5	0.5	0.5	0.6	.	1	1	1.7
Ukraine	1145	1145	1153	1153	1102	1059	1112	1094	1090	1065	1097	989	830	700	568	1094	1094	1094
United Kingdom	2319	2259	2244	2254	2256	2353	2438	2558	2644	2729	2702	2603	2514	2339	2219	2000	1842	1860
United States	21120	.	.	.	20855	20568	20168	20147	20899	20925	21040	20679	20846	21229	21423 f	20737	19125	20071
Yugoslavia g	47	50	50	53	58	58	58	60	63	62	66	57	49	54	52	88	115	147
European Community	.	.	.	.	.	10428 h	.	.	.	.	.	.	.	.	.	7300 i *	.	.

Source: data submitted by the Parties to the Convention on Long-Range Transboundary Air Pollution.

Emissions of nitrogen oxides: Notes

  * Projections based on current reduction plans.
  a Preliminary data.
  b Figures apply to the Federal Republic of Germany as prior to 1989.
  c Based on the IPCC-methodology.
  d No information as to whether nature is included.
  e Figures apply to the European part within EMEP.
  f Adopted from the EPA report: EPA-454/R-95-011 (1995).
  g Emissions from stationary sources only.
  h CORINAIR total of 12 EC member States.
  i 5th Community Programme of Policy and Action in Relation to the Environment and Sustainable Development.

Table 3. Emissions of ammonia (1980-2010*) in the ECE region
(thousands of tonnes of NH$_3$ per year)

	1980	1981	1982	1983	1984	1985	1986	1987	1988	1989	1990	1991	1992	1993	1994	2000*	2005*	2010*
Austria											91	91	92	93	93 a			
Belarus											4 b				4 c			
Belgium						74					79	78	77	80				
Bosnia and Herzegovina																		
Bulgaria											323	280	230	219	146	143	140	140
Canada																		
Croatia											37	31	27	25	24			
Cyprus																		
Czech Republic											105	88	77	97	92			
Denmark						152		152		146	140	134		126	126	103		
Finland						43		45			41		41			32	23	23
France											700	690 a	676 a	666 a				
Germany	572 d	557 d	564 d	576 d	585 d	588 d	580 d	572 d	560 d	556 d	759	670	649	634 a	622 a			
Germany, former GDR	263	264	253	265	268	269	266	265	270	262								
Greece																		
Hungary	170					170	170	150		170	176	150	140	140		170	160	150
Iceland																		
Ireland											126 b	126 b	126 b	126 b		126	126	
Italy						508 e	505 e	506 e	501 e	497 e	384 e							
Latvia																		
Liechtenstein	0.14				0.15						0.15					0.15	0.15	
Lithuania																		
Luxembourg											7			7	8	6		
Netherlands	234	240	244	244	246	256	258	258	237	232	236	237	188	197	172 a	82		50
Norway									39	39	39	40	41	40	41			
Poland	550					550	550	550	550	550	508	443	342	382	384			
Portugal											93	93	93	93	92			
Republic of Moldova																		
Romania																		

Table 3 (*continued*)

	1980	1981	1982	1983	1984	1985	1986	1987	1988	1989	1990	1991	1992	1993	1994	2000*	2005*	2010*
Russian Federation f	1189 g	1192 g	1214 g	1245 g	1247 g	1239 g	1286	1277	1269	1258	1191	1161	1084	903	772	.	.	.
Slovakia	.	.	.	.	.	.	.	.	.	.	62 b	.	61 b	.	47	.	.	.
Slovenia	.	.	.	.	.	.	.	.	.	.	27	.	.	.	.	27	27	27
Spain	.	.	.	.	.	.	.	.	.	.	353	354	352	345	.	55	54	53
Sweden	.	.	.	.	.	.	.	.	.	.	61	60	59	58	60	55	54	53
Switzerland	71	.	.	.	.	64	.	.	.	.	62	61	61	60	60	59	59	58
Turkey	.	.	.	.	.	.	.	.	.	.	.	.	.	.	.	.	.	.
Ukraine	.	.	.	.	.	.	.	.	.	.	.	.	.	.	.	.	.	.
United Kingdom	.	.	.	.	.	.	.	.	.	.	320	.	.	320	320	320	.	.
United States	.	.	.	.	.	1685 b	.	.	.	.	.	.	.	.	.	.	.	.
Yugoslavia	.	.	.	.	.	.	.	.	.	.	.	.	.	.	.	.	.	.
European Community	.	.	.	.	.	.	.	.	.	.	.	.	.	.	.	.	.	.

Source: data submitted by the Parties to the Convention on Long-Range Transboundary Air Pollution.

Emissions of ammonia: Notes

* Projections based on current reduction plans.
a   Preliminary data.
b   No information as to whether nature is included.
c   Emissions from stationary sources only.
d   Figures apply to the Federal Republic of Germany as prior to 1989.
e   Preliminary data. No information as to whether nature is included.
f   Figures apply to the European part within EMEP.
g   Agricultural sector only.

Table 4. Emissions of NMVOCs (1980-2010*) in the ECE region
(thousands of tonnes of HC per year)

	1980	1981	1982	1983	1984	1985	1986	1987	1988	1989	1990	1991	1992	1993	1994	2000*	2005*	2010*
Austria	374	-	-	391	-	412	-	439	432	434	430	419	403	388	364 a	305	-	-
Belarus	549	546	543	543	540	516	506	509	535	511	533	546	412	372	76 b	380	323	-
Belgium	-	-	-	-	-	660 c	-	-	-	-	365	363	369	362	-	-	-	-
Bosnia and Herzegovina	-	-	-	-	-	-	-	-	-	-	-	-	-	-	-	-	-	-
Bulgaria	-	-	-	-	-	-	-	-	-	-	217	204	199	199	205	356	276	265
Canada	2099	-	-	-	-	2851	2859	2897	2964	2906	2880	2792	2730	2763	2752	2694	2790	2927
Croatia	-	-	-	-	-	-	-	-	-	-	105	84	63	68	73	-	-	-
Cyprus	-	-	-	-	-	-	-	-	-	-	-	-	-	-	-	-	-	-
Czech Republic	-	-	-	-	-	275 d	-	-	-	-	534 a	497 a	459 a	438 a	399 a	-	-	-
Denmark	-	-	-	-	-	198	-	-	-	-	-	-	-	-	161	136	-	-
Finland	-	-	-	-	-	-	-	210	213	-	209	-	-	195	-	151	108	108
France	-	-	-	-	-	-	-	-	-	-	2404	2359 a	2329 a	2282 a	-	-	-	-
Germany	2614 e	2506 e	2539 e	2525 e	2537 e	2510 e	2501 e	2460 e	2401 e	2319 e	2985	2824	2753	2545 a	2405 a	-	1750	-
Germany, former GDR	682	684	649	663	684	710	722	748	766	782	-	-	-	-	-	-	-	-
Greece	-	-	-	-	-	614 c	-	-	-	-	-	-	-	-	-	-	-	-
Hungary	215	-	-	-	-	232	263	228	205	205	205	144	136	143	143 a	180	160	145
Iceland	-	-	-	-	-	-	-	6 f	6 f	6 f	6 f	7 f	7 f	6 f	6 f	6 f	6 f	6 f
Ireland	-	-	-	-	-	-	-	-	-	-	197 d	200 d	199 d	202 d	-	171	138	138
Italy	-	-	-	-	-	1771 g	1798 g	1865 g	1879 g	1913 g	2554 g	-	-	-	-	-	-	-
Latvia	-	-	-	-	-	-	-	-	-	-	-	-	-	-	-	-	-	-
Liechtenstein	1.48	1.49	1.49	1.50	1.51	1.52	1.53	1.53	1.54	1.55	1.56	1.49	1.43	1.36	1.30	0.92	0.86	-
Lithuania	-	-	-	-	-	20 c	-	-	-	-	19	-	-	-	-	-	-	-
Luxembourg	-	-	-	-	-	-	-	-	-	-	-	-	-	16	17	14	-	-
Netherlands	-	-	-	-	-	487	-	-	-	-	444	414	399	394	391	196	-	120
Norway	174	187	197	211	224	234	251	253	249	268	266	266	279	283	295	182	-	-
Poland	1036	912	889	954	985	1011	1029	1014	1026	1016	831	833	805	756	819	-	-	-
Portugal	-	-	-	-	-	199	-	-	-	-	206	209	219	222	231	-	-	-
Republic of Moldova	-	-	-	-	-	-	-	-	-	-	-	-	-	-	-	-	-	-
Romania	-	-	-	-	-	-	-	-	-	-	-	-	109 d	-	-	-	-	-

Table 4 (continued)

	1980	1981	1982	1983	1984	1985	1986	1987	1988	1989	1990	1991	1992	1993	1994	2000*	2005*	2010*
Russian Federation h	2843 d	2843 d	2582 d	2444 d	2390 d	2496 d	2338 d	2807 d	2790 d	3715 d	3566 d	3259 d	3204 d	2979 d	2861	·	·	·
Slovakia	·	·	·	·	·	·	·	·	·	·	149	·	124	107	108	·	·	·
Slovenia	·	·	·	·	·	·	·	·	39	·	35	·	·	·	·	33	30	25
Spain	·	·	·	·	·	1265	882	922	955	·	1134	1187	1207	1196	·	·	·	·
Sweden	·	·	·	·	·	·	·	·	555 d	·	528 d	·	502 d	·	·	342	287	·
Switzerland	323	·	·	·	324	324	·	·	·	·	292	274	256	239	226	172	170	173
Turkey	·	·	·	·	·	·	·	·	·	·	·	·	·	·	·	·	·	·
Ukraine	·	·	·	·	·	1626	1660	1687	1604	1512	1369	1302	1171	972	1024	1369	1369	1369
United Kingdom	2093	2101	2123	2117	2132	2165	2205	2240	2278	2324	2287	2269	2205	2110	2042	1519	1340	1276
United States	25719	24044	22556	23053	23014	22875	22343	21904	22465	21358	21087	21155	20438	20604	21023 i	19666	18416	19116
Yugoslavia	·	·	·	·	·	·	·	·	·	·	·	·	·	·	·	·	·	·
European Community	·	·	·	·	·	·	·	·	·	·	·	·	·	·	·	·	·	·

Source: data submitted by the Parties to the Convention on Long-Range Transboundary Air Pollution.

Emissions of NMVOCs: Notes

* Projections based on current reduction plans.
a Preliminary data.
b Emissions from stationary sources only.
c Including CH_4.
d No information as to whether nature is included.
e Figures apply to the Federal Republic of Germany as prior to 1989.
f Based on the IPCC-methodology.
g Preliminary data. No information as to whether nature is included.
h Figures apply to the European part within EMEP.
i Adopted from the EPA report: EPA-454/R-95-011 (1995).

Table 5. Emissions of methane (1980-2010*) in the ECE region
(thousands of tonnes of CH$_4$ per year)

	1980	1981	1982	1983	1984	1985	1986	1987	1988	1989	1990	1991	1992	1993	1994	2000*	2005*	2010*
Austria	.	.	.	.	.	.	.	.	.	.	603	606	608	608	608 a	.	.	.
Belarus	.	.	.	.	.	.	.	.	.	.	.	.	.	.	.	.	.	.
Belgium	.	.	.	.	.	.	.	.	.	.	355	370	407	534	.	.	.	.
Bosnia and Herzegovina	.	.	.	.	.	.	.	.	.	.	.	.	.	.	.	.	.	.
Bulgaria	.	.	.	.	.	.	.	646	660	638	757	612	624	664	523	451	420	420
Canada	2072	2682	2739	2759	2785	2847	2691	2801	2934	2987	3027	3102	3236	3364	.	3481	.	.
Croatia	.	.	.	.	.	.	.	.	.	.	169	151	136	147	138	.	.	.
Cyprus	.	.	.	.	.	.	.	.	.	.	.	.	.	.	.	.	.	.
Czech Republic	.	.	.	.	.	.	.	.	.	.	1455	1327	1224	1031	930	.	.	.
Denmark	403	405	406	406	406	407	407	407	407	406	406	408	408	407	367	355	354	.
Finland	.	.	.	.	.	.	.	.	520 b	.	249	.	.	.	.	.	.	.
France	.	.	.	.	.	.	.	.	.	.	2847	2808 a	2770 a	2768 a	.	.	.	.
Germany	4922 c	4804 c	4694 c	4600 c	4539 c	4607 c	4582 c	4455 c	4347 c	4344 c	5690	5272	5224	5229 a	5238 a	.	3250	.
Germany, former GDR	1196	1218	1223	1241	1266	1310	1295	1298	1307	1288	.	.	.	.	.	.	.	.
Greece	.	.	.	.	.	.	.	.	.	.	.	.	.	.	.	.	.	.
Hungary	.	.	.	.	.	706 b	.	628 b	.	.	581 b	.	.	.	.	330	.	.
Iceland	.	.	.	.	.	.	.	23 d	23 d	23 d	23 d	23 d	21 d	21 d	21 d	22 d	21 d	21 d
Ireland	.	.	.	.	.	.	.	.	.	.	800 b	800 b	800 b	800 b	.	800	.	.
Italy	.	.	.	.	.	2264 e	2331 e	2397 e	2469 e	2543 e	3778 b	.	.	.	.	.	.	.
Latvia	.	.	.	.	.	.	.	.	.	.	.	.	.	.	.	.	.	.
Liechtenstein	0.74	.	.	.	0.71	.	.	.	.	.	0.70	.	.	.	.	0.58	0.50	.
Lithuania	.	.	.	.	.	.	.	.	.	.	.	.	.	.	.	.	.	.
Luxembourg	.	.	.	.	.	.	.	.	.	.	24	.	.	24	24	25	.	.
Netherlands	971	875	876	1009	1039	1049	1049	1047	1024	1039	1060	1090	1072	1067	1041	1091	.	.
Norway	264	.	.	.	.	.	.	281	281	287	290	289	293	290	293	.	.	.
Poland	.	.	.	.	.	.	.	.	.	.	5715	.	3647 b	2990	1841	.	.	.
Portugal	.	.	.	.	.	.	.	.	.	.	254	254	254	254	252	.	.	.
Republic of Moldova	.	.	.	.	.	.	.	.	.	.	.	.	.	.	.	.	.	.
Romania	.	.	.	.	.	.	.	.	.	.	.	.	1424 b	.	.	.	.	.

Table 5 (continued)

	1980	1981	1982	1983	1984	1985	1986	1987	1988	1989	1990	1991	1992	1993	1994	2000*	2005*	2010*
Russian Federation i	5462 b	5414 b	5396 b	5423 b	5441 b	5427 b	12900 g	5428 b	5358 b	5312 b	5174 b	4923 b	4649 b	3680 b	2942	.	.	.
Slovakia	.	.	.	.	.	.	.	.	.	.	354	.	337 h	.	290	.	.	.
Slovenia	.	.	.	.	.	.	.	.	90	.	124	.	.	.	.	119	100	87
Spain	.	.	.	.	.	.	.	.	.	.	2185	2166	2259	2310	.	.	.	.
Sweden	.	.	.	.	.	.	.	.	.	.	340 b	339 b	337 b	336 b	.	300	296	.
Switzerland	435	.	.	.	.	385	.	.	.	.	332	327	323	318	316	300	284	266
Turkey	.	.	.	.	.	.	.	.	.	.	.	.	.	.	.	.	.	.
Ukraine	.	.	.	.	.	.	.	.	.	.	.	.	.	.	.	.	.	.
United Kingdom	4797	4772	4760	4734	4036	4500	4653	4596	4538	4487	4429	4391	4280	4107	3879	4257	3993	3730
United States	.	.	.	.	.	.	.	.	.	.	27550 l	.	.	.	.	.	.	.
Yugoslavia	.	.	.	.	.	.	.	.	.	.	.	.	.	.	.	.	.	.
European Community	.	.	.	.	.	.	.	.	.	.	.	.	.	.	.	.	.	.

Source: data submitted by the Parties to the Convention on Long-Range Transboundary Air Pollution.

Emissions of methane: Notes

* Projections based on current reduction plans.
a Preliminary data.
b No information as to whether nature is included.
c Figures apply to the Federal Republic of Germany as prior to 1989.
d Based on the IPCC-methodology.
e Preliminary data. No information as to whether nature is included.
f Figures apply to the European part within EMEP.
g Preliminary data including emissions from hydrocarbon extraction. No information as to whether nature is included
h Preliminary data including nature.
i 20400 to 34700 kt.

Table 6. Emissions of carbon monoxide (1980-2010*) in the ECE region
(thousands of tonnes of CO per year)

	1980	1981	1982	1983	1984	1985	1986	1987	1988	1989	1990	1991	1992	1993	1994	2000*	2005*	2010*
Austria	1636	.	.	1561	.	1648	.	1685	1578	1605	1573	1503	1414	1326	1408a	.	.	.
Belarus	.	.	.	.	.	.	.	.	.	.	191b	.	.	.	110c	.	.	.
Belgium	.	.	.	.	.	.	.	.	.	.	1124	1131	1177	1147	.	.	.	.
Bosnia and Herzegovina	.	.	.	.	.	.	.	.	.	.	.	.	.	.	.	.	.	.
Bulgaria	.	.	.	.	.	.	.	997	995	985	891	654	764	776	1008	820	800	750
Canada	10273	.	.	.	.	9685	.	.	.	.	10596	10153	9855	9851	9747	9182	9809	10550
Croatia	.	.	.	.	.	.	.	.	.	.	651	564	430	421	466	.	.	.
Cyprus	.	.	.	.	.	.	.	.	.	.	.	.	.	.	.	.	.	.
Czech Republic	894	.	906	.	895	899	740	738	737	884	1055	1102	1045	967	978	.	.	.
Denmark	673	675	686	702	727	741	753	754	756	744	770	824	812	732	728	647	562	.
Finland	.	.	.	.	.	.	.	.	.	.	556b	.	.	.	.	.	.	.
France	9216	9146	8858	8648	8529	8399	8156	8036	7821	7575	10736	10599a	10265a	9755a	.	.	.	.
Germany	12014d	10773d	9975d	9294d	9324d	8897d	8726d	8382d	8085d	7667d	10280	9032	8640	8029a	7428a	.	5350	.
Germany, former GDR	3050	3005	2943	2840	2941	3152	3211	3326	3215	3094	.	.	.	.	.	.	.	.
Greece	.	.	.	.	.	.	.	.	.	.	.	.	.	.	.	.	.	.
Hungary	.	.	.	.	.	.	.	.	.	.	804	914	836	838	838a	800	800	800
Iceland	.	.	.	.	.	.	.	24e	26e	26e	26e	26e	26e	25e	25e	19e	18e	18e
Ireland	.	.	.	.	.	.	.	.	.	.	429b	428b	403b	416b	.	322	322	322
Italy	.	.	.	.	.	6919b	6821b	6744b	6668b	6591b	10347b	.	.	.	.	.	.	.
Latvia	.	.	.	.	.	.	.	.	.	.	.	.	.	.	.	.	.	.
Liechtenstein	4.19	4.00	3.80	3.61	3.42	3.23	3.04	2.85	2.66	2.47	2.27	2.16	2.04	1.92	1.80	1.10	1.05	.
Lithuania	.	.	.	.	.	.	.	.	.	.	.	.	.	.	.	.	.	.
Luxembourg	.	.	.	.	.	240	.	.	.	.	171	.	.	219	142	102	.	.
Netherlands	.	.	.	.	.	1356	.	.	.	.	1059	959	941	917	897a	.	.	.
Norway	886	873	894	897	926	962	1007	1014	997	954	940	881	849	805	789	.	.	.
Poland	.	.	.	.	.	.	.	.	.	.	7406b	.	7083b	8655	5115	.	.	.
Portugal	.	.	.	.	.	.	.	.	.	.	1086	1111	1156	1175	1211	.	.	.
Republic of Moldova	.	.	.	.	.	.	.	.	.	.	.	.	.	.	.	.	.	.
Romania	.	.	.	.	.	.	.	1016b	1142b	2655b	2098b	2376b	498b	.	.	.	.	.

Table 6 (continued)

	1980	1981	1982	1983	1984	1985	1986	1987	1988	1989	1990	1991	1992	1993	1994	2000*	2005*	2010*
Russian Federation f	13512b	15005b	13617b	13696b	13672b	14122b	13142b	13119b	12988b	13054b	13174b	12869b	11574b	11193b	10495	.	.	.
Slovakia	.	.	.	.	.	.	.	.	.	.	462b	.	343b	.	411	.	.	.
Slovenia	.	.	.	.	.	.	.	.	.	.	77	.	.	.	93	57	44	31
Spain	.	.	.	.	.	.	.	.	.	.	4752	4822	4787	4801	.	.	.	.
Sweden	.	.	.	.	.	.	.	.	.	.	1347b	1312b	1275b	1236b	.	760	631	370
Switzerland	1280	.	.	.	.	990	.	.	.	.	707	665	621	578	549	408	369	.
Turkey	.	.	.	.	.	.	.	.	.	.	.	.	.	.	.	.	.	.
Ukraine	.	.	.	.	.	9832	9722	9269	9085	8794	8141	7406	5496	4218	3375	8141	8141	8141
United Kingdom	5631	5705	5853	5756	5810	5895	6006	6161	6346	6588	6360	6287	5842	5312	4884	3324	1884	1374
United States	117032	111583	105369	105200	102835	100865	97263	94605	95490	90725	87295	85541	84178	84203	88920g	81461	74790	78452
Yugoslavia	.	.	.	.	.	.	.	.	.	.	.	.	.	.	.	.	.	.
European Community	.	.	.	.	.	.	.	.	.	.	.	.	.	.	.	.	.	.

Source: data submitted by the Parties to the Convention on Long-Range Transboundary Air Pollution.

Emissions of carbon monoxide: Notes

* Projections based on current reduction plans.
a Preliminary data.
b No information as to whether nature is included.
c Stationary sources only.
d Figures apply to the Federal Republic of Germany as prior to 1989.
e Based on the IPCC-methodology.
f Figures apply to the European part within EMEP.
g Adopted from the EPA report: EPA-454/R-95-011 (1995).

Table 7. Emissions of carbon dioxide (1980-2010*) in the ECE region
(millions of tonnes of CO_2 per year)

	1980	1981	1982	1983	1984	1985	1986	1987	1988	1989	1990	1991	1992	1993	1994	2000*	2005*	2010*
Austria	59 a	.	.	56 a	.	57 a	.	58 a	56 a	57 a	60 a	64 a	59 a	59 a	60 a	66 b	.	.
Belarus	.	.	.	.	.	.	.	.	.	.	.	.	.	.	.	.	.	.
Belgium	.	.	.	.	.	.	.	.	.	.	.	.	.	101	.	.	.	.
Bosnia and Herzegovina	.	.	.	.	.	.	.	.	.	.	.	.	.	.	.	.	.	.
Bulgaria	.	.	.	.	.	.	.	100 a	102 a	98 a	91 a	68 a	68 a	72 a	75 a	80 a	80 a	80 a
Canada	440 c	442 c	401 c	387 c	404 c	421 c	416 c	434 c	466 c	490 c	460 c	452 c	466 c	470 c	.	518 c	546 c	564 c
Croatia	.	.	.	.	.	.	.	.	.	.	24	18	16	17	18	26	30	33
Cyprus	.	.	.	.	.	4 d	.	.	.	.	6 d	5 d	6 d	6 d	7 d	9	12	14
Czech Republic	196	.	.	.	.	180	179	178	173	170	164	155	142	138	130	144	153	.
Denmark	63	53	55	53	54	62	62	60	56	50	52	63	57	58	63	54	52	.
Finland	.	.	.	.	.	.	.	.	.	.	55 e	.	68 f	71 f	78 f	65 e	65 e	65 e
France	503	453	432	413	403	388	374	368	371	384	373	399 a	388 a	379 a	.	.	.	.
Germany	798 g	762 g	728 g	728 g	739 g	731 g	736 g	724 g	716 g	702 g	1027	988	940	922 a	912 a	.	.	.
Germany, former GDR	324	325	323	320	338	354	356	358	353	349	.	.	.	.	.	.	.	.
Greece	48 d	.	.	.	.	59 d	58 d	63 d	67 d	72 d	.	.	.	.	.	.	.	.
Hungary	92	92	92	91	90	89	87	87	84	81	74	72	66	66	64 a	71	75	78
Iceland	.	.	.	.	.	.	.	2 h	2 h	2 h	2 h	2 h	2 h	2 h	2 h	2 h	2 h	2 h
Ireland	.	.	.	.	.	.	.	.	.	.	31 d	31 d	31 d	32 d	.	37	.	.
Italy	.	.	.	.	.	408 d	414 d	434 d	443 d	458 d	441 d	.	.	.	.	232	.	.
Latvia	.	.	.	.	.	.	.	.	.	.	.	.	.	.	.	.	.	.
Liechtenstein	0.15	0.15	0.13	0.13	0.15	0.18	0.17	0.18	0.18	0.20	0.21	0.21	0.22	0.22	0.22	0.25	0.26	.
Lithuania	.	.	.	.	.	.	.	.	.	.	.	.	.	.	.	.	.	.
Luxembourg	.	.	.	.	.	10	.	.	.	.	11	.	.	11	8	6	.	.
Netherlands	167	.	.	.	.	152	.	.	.	.	168	174	172	174	176	175 i	.	.
Norway	34	31	30	31	33	32	34	35	35	35	36	34	34	36	37	35	.	.
Poland	.	.	.	.	.	.	.	.	509 d	488 d	407 d	397 d	393 d	348 d	348	.	.	.
Portugal	.	.	.	.	.	.	.	.	.	.	47	49	53	51	51	54 j	.	.
Republic of Moldova	.	.	.	.	.	.	.	.	.	.	.	.	.	.	.	.	.	.
Romania	.	.	.	.	.	.	.	134 d	127 d	132 d	130 d	106 d	198 d	.	.	.	.	.

Table 7 (continued)

	1980	1981	1982	1983	1984	1985	1986	1987	1988	1989	1990	1991	1992	1993	1994	2000 *	2005 *	2010 *
Russian Federation k	.	.	.	.	.	1560 d	.	.	1650 d	.	1670 d	1630 d	1630 d	.	1580	1760	.	1900
Slovakia	.	.	.	.	.	.	.	.	58 d	.	55 a	.	50 a	.	.	.	.	.
Slovenia	.	.	.	.	.	.	.	.	14	14	13	12	13	13	14	12	12	11
Spain	.	.	.	.	.	205	199	197	185	217	203 h	204 h	211 h	203 h	208 h	231	.	.
Sweden	82 l	74 l	69 l	64 l	63 l	67 l	68 l	67 l	63 l	64 l	60 l	59 l	59 l	61 l	63 d	63	67	121
Switzerland	.	.	.	.	.	.	.	.	.	.	45 m	.	.	.	.	46 n	.	.
Turkey	76 d	76 d	88 d	88 d	94 d	102 d	112 d	120 d	115 d	128 d	137 d	141 d	147 d	153 d	.	247	316	435
Ukraine	.	.	.	.	.	.	.	.	.	.	.	.	.	.	.	.	.	.
United Kingdom	163	157	153	151	147	153	157	158	158	155	157	159	154	150	149	168	164	.
United States	.	.	.	.	.	.	.	.	.	.	4400	.	.	.	.	.	.	.
Yugoslavia	.	.	.	.	.	.	.	.	.	.	.	.	.	.	.	.	.	.
European Community	2850	2715	2620	2575	2605	2660	2700	2600	..	2765	.	.	.	.	.	2765	.	.

Source: data submitted by the Parties to the Convention on Long-Range Transboundary Air Pollution.

Emissions of carbon dioxide: Notes

  * Projections based on current reduction plans.
a Preliminary data.
b Excluding emissions from biogenic fuels.
c Including emissions from marine air bunkers.
d No information as to whether nature is included.
e From fuels.
f CO_2 emissions from fossil fuels and peat for 1992, 1993 and 1994 are 51, 51 and 57 million tonnes, respectively.
g Figures apply to the Federal Republic of Germany as prior to 1989.
h Based on the IPCC-methodology.
i 173-177 kt.
j Emissions from combustion only.
k Figures apply to the European part within EMEP.
l Excluding biomass fuels and biomass related wastes.
m Without climatic correction.
n With climatic correction.

Table 8.   Convention on Long-range Transboundary Air Pollution and its related Protocols (as of 31 March 1996)

	1979 Convention (a)		1984 EMEP Protocol (b)		1985 Sulphur Protocol (c)		1988 NOx Protocol (d)		1991 VOC Protocol (e)		1994 Sulphur Protocol (f)	
	Signature	Ratification*	Signature	Ratification*	Signature	Ratification*	Signature	Ratification*	Signature (1)	Ratification (2)	Signature	Ratification*
Austria	13.11.1979	16.12.1982 (R)		04.06.1987 (Ac)	9.7.1985	04.06.1987 (R)	1.11.1988	15.01.1990 (R)	19.11.1991	23.08.1994 (R)	14.06.1994	
Belarus	14.11.1979	13.06.1980 (R)	28.09.1984	04.10.1985 (AI)	9.7.1985	10.09.1986 (AI)	1.11.1988	08.06.1989 (AI)				
Belgium	13.11.1979	15.07.1982 (R)	25.02.1985	05.08.1987 (R)	9.7.1985	09.06.1989 (R)	1.11.1988		19.11.1991		14.06.1994 (1)	
Bosnia and Herzegovina												
Bulgaria	14.11.1979	09.06.1981 (R)	04.04.1985	26.09.1986 (Ap)	9.7.1985	26.09.1986 (Ap)	1.11.1988	30.03.1989 (R)	19.11.1991		14.06.1994	
Canada	13.11.1979	15.12.1981 (R)	03.10.1984	04.12.1985 (R)	9.7.1985	04.12.1985 (R)	1.11.1988	25.01.1991 (R)	19.11.1991		14.06.1994	
Croatia		08.10.1992(Sc)		03.10.1992(Sc)								
Cyprus		20.11.1991 (Ac)		23.11.1991 (Ac)								
Czech Republic		01.01.1993 (Sc)		01.01.1993 (Sc)		01.01.1993 (Sc)		01.01.1993 (Sc)			14.06.1994	
Denmark	14.11.1979	18.06.1982 (R)	28.09.1984	23.04.1986 (R)	9.7.1985	29.04.1986 (R)	1.11.1988	01.03.1993 (AI)(2)	19.11.1991	11.01.1994 (AI)	14.06.1994	
Finland	13.11.1979	15.04.1981 (R)	07.12.1984	21.06.1986 (R)	9.7.1985	24.06.1986 (R)	1.11.1988	01.02.1990 (R)	19.11.1991		14.06.1994	
France	13.11.1979	03.11.1981 (Ap)	22.02.1985	30.10.1987 (R)	9.7.1985	13.03.1986 (Ap)	1.11.1988	20.07.1989 (Ap)	19.11.1991		14.06.1994	
Germany	13.11.1979	15.07.1982 (R)(2)	26.02.1985	07.10.1986 (R)(2)	9.7.1985	03.03.1987 (R)(2)	1.11.1988	16.11.1990 (R)	19.11.1991	08.12.1994 (R)	14.06.1994	
Greece	14.11.1979	30.08.1983 (R)		24.06.1988 (Ac)					19.11.1991		14.06.1994	
Holy See	14.11.1979											
Hungary	13.11.1979	22.09.1980 (R)	27.03.1985	08.05.1985 (Ap)	9.7.1985	11.09.1986 (R)	3.05.1989	12.11.1991 (Ap)	19.11.1991	10.11.1995 (R)	09.12.1994	
Iceland	13.11.1979	05.05.1983 (R)					1.05.1989	17.10.1994 (R)			17.10.1994	
Ireland	13.11.1979	15.07.1982 (R)	04.04.1985	26.06.1987 (R)								
Italy	14.11.1979	15.07.1982 (R)	28.09.1984	12.01.1989 (R)	9.7.1985	05.02.1990 (R)	1.11.1988	19.05.1992 (R)	19.11.1991	30.06.1995 (R)	14.06.1994	
Latvia		15.07.1994 (Ac)										
Liechtenstein	14.11.1979	22.11.1983 (R)		01.05.1985 (Ac)	9.7.1985	13.02.1986 (R)	1.11.1988	24.03.1994 (R)	19.11.1991	24.03.1994 (R)	14.06.1994	
Lithuania		25.01.1994 (Ac)										
Luxembourg	13.11.1979	15.07.1982 (R)	21.11.1984	24.08.1987 (R)	9.7.1985	24.08.1987 (R)	1.11.1988	04.10.1990 (R)	19.11.1991	11.11.1993 (R)	14.06.1994	30.05.1995 (AI)(2)
Netherlands	13.11.1979	15.07.1982 (AI)(3)	28.09.1984	22.10.1985 (AI)(3)	9.7.1985	30.04.1986 (AI)(3)	1.11.1988	11.10.1989 (AI)(3)	19.11.1991	29.09.1993 (AI)	14.06.1994	30.05.1995 (AI)(2)
Norway	13.11.1979	13.02.1981 (R)	28.09.1984	12.03.1985 (AI)	9.7.1985	04.11.1986 (R)	1.11.1988	11.10.1989 (R)	19.11.1991	07.01.1993 (R)	14.06.1994	03.07.1995 (R)
Poland	13.11.1979	19.07.1985 (R)(2)		14.09.1988 (Ac)			1.11.1988					
Portugal	14.11.1979	29.09.1980 (R)		10.01.1989 (Ac)					02.04.1992			
Republic of Moldova		09.06.1995 (Ac)										
Romania	14.11.1979 (1)	27.02.1991 (R)										
Russian Federation	13.11.1979	22.05.1980 (R)	28.09.1984	21.08.1985 (AI)	9.7.1985	10.09.1986 (AI)	1.11.1988	21.06.1989 (AI)			14.06.1994	
San Marino	14.11.1979											
Slovakia		28.05.1993 (Sc)		28.05.1993 (Sc)		28.05.1993 (Sc)		28.05.1993 (Sc)			14.06.1994	
Slovenia		06.07.1992 (Sc)		06.07.1992 (Sc)							14.06.1994	
Spain	14.11.1979	15.06.1982 (R)		11.08.1987 (Ac)			1.11.1988	04.12.1990 (R)	19.11.1991	01.02.1994 (R)	14.06.1994	
Sweden	13.11.1979	12.02.1981 (R)	28.09.1984	12.08.1985 (R)	9.7.1985	31.03.1986 (R)	1.11.1988	27.07.1990 (R)	19.11.1991	08.01.1993 (R)	14.06.1994	19.07.1995 (R)
Switzerland	13.11.1979	06.05.1983 (R)	03.10.1984	26.07.1985 (R)	9.7.1985	21.09.1987 (R)	1.11.1988	18.09.1990 (R)	19.11.1991	21.03.1994 (R)	14.06.1994	
Turkey	13.11.1979	18.04.1983 (R)	03.10.1984	20.12.1985 (R)			1.11.1988		19.11.1991			
Ukraine	14.11.1979	05.06.1980 (R)	28.09.1984	30.08.1985 (AI)	9.7.1985	02.10.1986 (AI)	1.11.1988	24.07.1989 (AI)	19.11.1991		14.06.1994	
United Kingdom	13.11.1979	15.07.1982 (R)(4)	20.11.1984	12.08.1985 (R)			1.11.1988	15.10.1990 (R)(4)	19.11.1991	14.06.1994 (R)(5)	14.06.1994	
United States	13.11.1979	30.11.1981 (AI)	28.09.1984	29.10.1984 (AI)			1.11.1988 (1)	13.07.1989 (AI)	19.11.1991			
Yugoslavia	14.11.1979	18.03.1987 (R)	28.09.1984	28.10.1987 (Ap)								
European Community	14.11.1979	15.07.1982 (Ap)	28.09.1984	17.07.1986 (Ap)				17.12.1993 (Ac)	02.04.1992	02.04.1992 (Ac)	14.06.1994	
Total:	33	40	22	35	19	21	25	25	23	13	28	3

(a) Convention on Long-range Transboundary Air Pollution, adopted 13.11.1979 in Geneva, entry into force 16.3.1983 (E/ECE/1010).

(b) Protocol to the 1979 Convention on Long-range Transboundary Air Pollution on Long-term Financing of the Cooperative Programme for Monitoring and Evaluation of the Long-range Transmission of Air Pollutants in Europe (EMEP), adopted 28.9.1984 in Geneva, entry into force 28.1.1988 (ECE/EB.AIR/11).

(c) Protocol to the 1979 Convention on Long-range Transboundary Air Pollution on the Reduction of Sulphur Emissions or their Transboundary Fluxes by at least 30 per cent, adopted 8.7.1985 in Helsinki, entry into force 2.9.1987 (ECE/EB.AIR/12).

(d) Protocol to the 1979 Convention on Long-range Transboundary Air Pollution concerning the Control of Emissions of Nitrogen Oxides or their Transboundary Fluxes, adopted 31.10.1988 in Sofia, entry into force 14.2.1991 (ECE/EB.AIR/21).

(e) Protocol to the 1979 Convention on Long-range Transboundary Air Pollution concerning the Control of Emissions of Volatile Organic Compounds or their Transboundary Fluxes, adopted 18.11.1991 in Geneva (ECE/EB.AIR/30).

(f) Protocol to the 1979 Convention on Long-range Transboundary Air Pollution on Further Reduction of Sulphur Emissions, adopted 14.6.1994 in Oslo (ECE/EB.AIR/40).

Notes: * R = Ratification, Ac = Accession, Ap = Approval, AI = Acceptance, Sc = Succession

(1) With declaration upon signature.   (†) Including the Bailiwicks of Jersey and Guernsey, the Isle of Man.

(2) With declaration upon ratification.   (3) For the Kingdom in Europe.   (4) Including the Bailiwicks of Jersey and Guernsey, the Isle of Man, Gibraltar, the United Kingdom Sovereign Base Areas of Akrotiri and Dhekelia on the island of Cyprus.

Table 9. Emissions of sulphur (1980-2010) in the ECE region as a percentage of 1980 levels

Country	1980	1981	1982	1983	1984	1985	1986	1987	1988	1989	1990	1991	1992	1993	1994	2000	2005	2010
Austria	100	-	-	61	-	49	-	38	31	23	23	21	19	18	19	20	-	-
Belarus	100	99	96	96	93	93	93	110	105	97	96	98	69	59	51	75	66	-
Belgium	100	86	84	68	60	48	46	44	43	39	38	39	37	36	-	30	28	26
Bosnia and Herzegovina	-	-	-	-	-	-	-	-	-	-	-	-	-	-	-	-	-	-
Bulgaria	100	-	-	-	-	-	-	118	109	106	99	81	55	69	72	67	60	55
Canada	100	92	78	78	85	80	78	81	83	80	70	70	67	65	57	61	63	63
Croatia	100	-	-	-	-	-	-	-	-	-	120	72	71	76	59	89	83	78
Cyprus	-	-	-	-	-	-	-	-	-	-	-	-	-	-	-	-	-	-
Czech Republic	100	104	106	104	102	101	96	96	92	89	83	79	68	63	56	50	40	28
Denmark	100	80	82	70	66	75	63	56	54	43	40	54	42	35	35	20	20	-
Finland	100	91	83	64	63	66	57	56	52	42	45	33	24	21	20	20	-	22
France	100	78	75	63	56	44	40	39	37	40	39	41	37	34	40	26	23	22
Germany	100	99	99	98	102	103	102	98	86	82	71	56	46	42	-	13	10	-
Greece	100	-	-	-	-	125	-	-	-	-	128	-	-	-	-	149	145	143
Hungary	100	97	95	91	88	86	83	79	75	67	62	56	51	46	45	55	50	40
Iceland	100	-	-	-	100	100	-	383	400	400	400	383	400	400	400	400	383	383
Ireland	100	86	71	64	64	63	73	78	68	73	80	81	73	71	-	70	-	-
Italy	100	-	-	83	70	59	59	60	58	53	44	43	40	39	-	32	27	-
Latvia	-	-	-	-	-	-	-	-	-	-	-	-	-	-	-	-	-	-
Liechtenstein	100	95	87	82	74	69	64	56	51	44	38	38	36	36	33	28	28	-
Lithuania	-	-	-	-	-	-	-	-	-	-	-	-	-	-	-	-	-	-
Luxembourg	100	-	-	58	-	71	-	-	-	-	58	-	-	63	50	17	-	-
Netherlands	100	95	82	66	61	53	54	54	51	42	42	40	35	33	30	19	-	11
Norway	100	90	78	73	67	69	65	52	48	42	38	32	26	26	25	24	-	-
Poland	100	-	-	-	-	105	102	102	102	95	78	73	69	66	64	63	53	34
Portugal	100	-	-	-	-	74	-	-	-	-	106	112	132	113	102	114	111	-
Republic of Moldova	-	-	-	-	-	-	-	-	-	-	-	-	-	-	-	-	-	-
Romania	-	-	-	-	-	-	-	-	-	-	-	-	-	-	-	-	-	-
Russian Federation	-	97	99	97	91	86	80	79	72	65	62	61	54	48	42	62	60	60
Slovakia	100	-	-	-	-	79	77	79	76	73	70	57	49	42	31	43	38	31
Slovenia	100	108	109	115	106	102	104	93	89	90	83	77	80	77	75	39	19	16
Spain	100	-	-	77	-	66	59	57	48	59	68	67	66	62	-	65	-	-
Sweden	100	85	73	60	58	52	54	45	44	31	27	22	20	20	19	20	-	-
Switzerland	100	-	-	-	72	66	-	-	-	-	37	35	33	29	27	26	26	26
Turkey	100	-	-	-	32	37	41	-	-	-	-	-	-	-	-	-	-	-
Ukraine	100	91	89	91	90	90	88	85	83	80	98	66	62	57	45	60	60	60
United Kingdom	100	91	86	79	76	76	80	79	78	76	77	73	71	65	55	47	30	20
United States	100	95	89	87	89	88	85	84	85	86	85	85	82	84	81	73	63	59
Yugoslavia	100	100	101	108	112	118	116	119	124	125	125	110	98	99	104	167	219	280
European Community	100	-	-	-	-	53	-	-	-	-	-	-	-	-	-	35	-	-

Notes: See footnotes to table 1.

Figure I.   Emissions of sulphur in the ECE region as a percentage of 1980 levels
(based on the latest data available, see table 1)

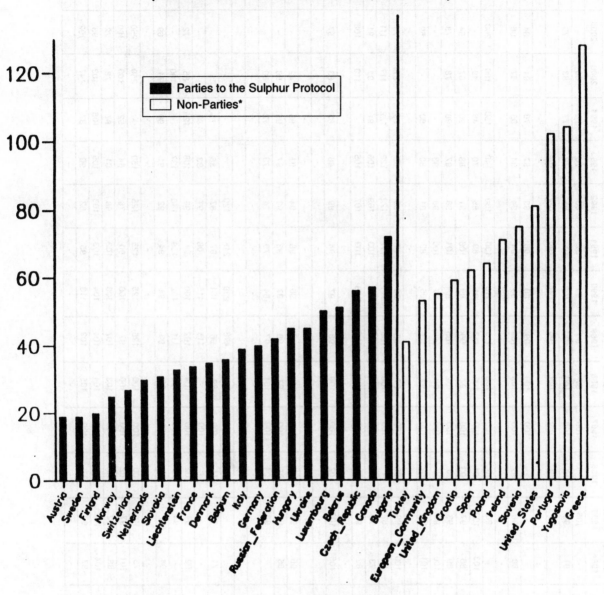

* For Bosnia and Herzegovina and Cyprus no emission data have been received for the reference year. For Iceland, no percentage is shown because the method of calculation for sulphur emissions changed in 1987.

Table 10. Emissions of nitrogen oxides (1980-2010) in the ECE region as a percentage of 1987 levels

Country	1980	1981	1982	1983	1984	1985	1986	1987	1988	1989	1990	1991	1992	1993	1994	2000	2005	2010
Austria	105	-	-	103	-	105	-	100	97	94	95	92	86	78	76	66	-	-
Belarus	89	89	89	90	91	90	98	100	100	100	108	107	85	79	77	83	70	-
Belgium	138	-	-	-	-	98	96	100	104	108	107	108	109	109	-	-	-	-
Bosnia and Herzegovina	-	-	-	-	-	-	-	-	-	-	-	-	-	-	-	-	-	-
Bulgaria	-	-	-	-	-	-	-	100	100	99	90	66	63	57	79	91	84	70
Canada	92	89	89	84	88	96	96	100	103	103	99	94	94	94	95	97	96	98
Croatia	-	-	-	-	-	-	-	-	-	-	-	-	-	-	-	-	-	-
Cyprus	-	-	-	-	-	-	-	100	100	110	110	130	130	140	150	180	200	200
Czech Republic	115	100	100	102	103	102	101	100	105	113	91	89	86	70	45	49	64	-
Denmark	91	79	86	84	88	97	103	100	97	90	89	106	91	88	90	67	-	-
Finland	102	96	94	91	89	95	96	100	102	105	104	101	98	97	98	78	78	78
France	112	104	104	101	100	99	99	100	99	109	97	100	98	93	-	-	-	-
Germany	102	99	98	99	101	101	102	100	97	94	85	82	81	80	80	-	-	-
Greece	-	-	-	-	-	-	-	-	-	-	-	-	-	-	-	-	59	-
Hungary	103	102	101	100	100	99	100	100	97	93	90	77	69	65	69	87	79	74
Iceland	72	-	-	-	67	67	87	100	106	106	111	117	122	128	122	117	111	106
Ireland	63	75	75	74	73	79	-	100	106	110	100	103	109	106	91	91	91	-
Italy	78	-	-	-	82	91	95	100	104	107	108	109	107	105	-	110	108	-
Latvia	-	-	-	-	-	-	-	-	-	-	-	-	-	-	-	-	-	-
Liechtenstein	109	109	108	106	105	103	102	100	100	98	97	94	89	86	83	63	57	-
Lithuania	-	-	-	-	-	-	-	-	-	-	-	-	-	-	-	-	-	-
Luxembourg	-	-	-	-	-	-	-	-	-	-	-	-	-	-	-	-	-	-
Netherlands	97	96	94	93	96	96	98	100	101	97	96	96	94	91	88	42	-	20
Norway	78	75	77	79	86	91	97	100	97	98	97	93	93	97	95	68	-	-
Poland	80	-	-	-	-	98	104	100	101	97	84	79	74	73	72	88	-	-
Portugal	-	-	-	-	-	-	-	-	-	-	-	-	-	-	-	-	-	-
Republic of Moldova	-	-	-	-	-	-	-	-	-	-	-	-	-	-	-	-	-	-
Romania	-	-	-	-	-	-	71	100	69	475	239	218	120	-	-	-	-	-
Russian Federation	65	72	75	74	71	72	-	100	89	96	101	97	87	86	75	-	-	-
Slovakia	-	-	-	-	-	-	-	100	89	115	115	108	97	93	88	-	-	-
Slovenia	91	92	92	91	92	94	102	100	104	102	100	94	96	108	125	85	72	58
Spain	107	-	-	105	-	94	96	100	100	111	132	138	140	137	-	100	-	-
Sweden	104	95	94	92	94	97	99	100	99	96	94	94	92	91	90	71	69	71
Switzerland	-	-	-	-	-	-	-	-	-	-	-	-	-	-	-	-	-	-
Turkey	75	75	75	75	75	-	100	100	100	125	125	125	125	150	-	250	250	425
Ukraine	105	105	105	105	101	97	102	100	100	97	100	90	76	64	52	100	100	100
United Kingdom	91	88	88	88	88	92	95	100	103	107	106	102	98	91	87	78	72	73
United States*	105	-	-	-	104	102	95	100	104	104	104	103	103	105	106	103	95	100
Yugoslavia	78	83	83	88	97	97	97	100	105	103	110	95	82	90	87	147	192	245
European Community	-	-	-	-	-	-	-	-	-	-	-	-	-	-	-	-	-	-

Notes: See footnotes to table 2.
* Reference year for the United States is 1978 (21830 kt).

Figure II. Emissions of nitrogen oxides in the ECE region as a percentage of 1987 levels (based on the latest data available, see table 2)

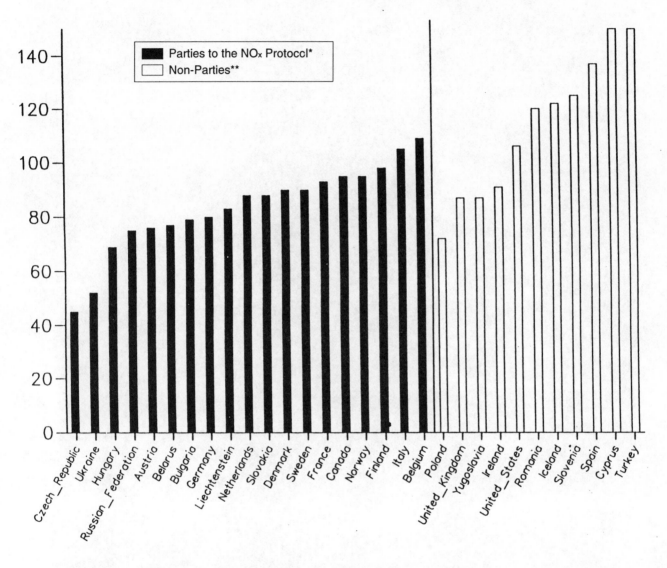

* For Luxembourg, Switzerland and the European Community no emission data have been received for the reference year.
** For Croatia, Greece and Portugal no emission data have been received for the reference year.

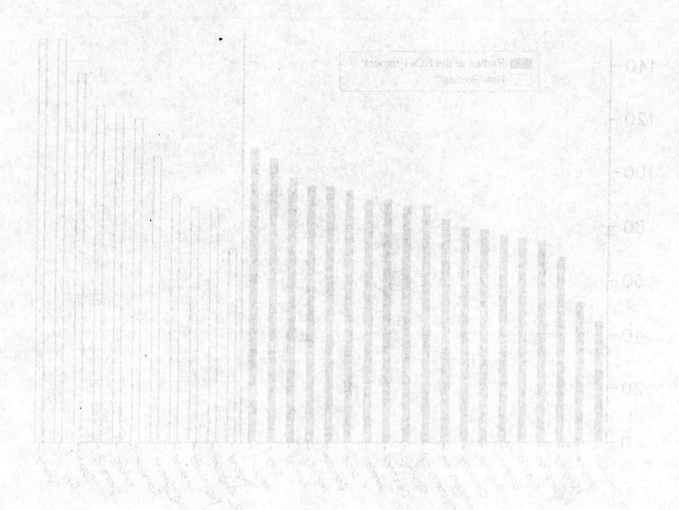

Figure II. Emissions of nitrogen oxides in the ECE region as a percentage of 1987 levels (based on the latest data available, see table 2)

Part TWO

FOREST CONDITION IN EUROPE: 1994 SURVEY

SUMMARY

The report describes the results of both the national and the transnational crown condition surveys, which are conducted annually within the International Cooperative Programme on the Assessment and Monitoring of Air Pollution Effects on Forests (ICP Forests) of the United Nations Economic Commission for Europe (ECE) and under European Union Council Regulation (EEC) 3528/86 on the Protection of the Community's Forests against Atmospheric Pollution. The report presents the survey results from 32 European countries, referring to 29,739 sample plots with 648,425 sample trees. The plots cover about 178.4 million hectares of forests. Twenty-nine of these countries have also submitted results from the 16 km x 16 km grid (transnational survey). The results of the 1994 survey indicate that forest damage continues to be a serious problem in Europe, as a substantial proportion of the trees was defoliated and/or discoloured. Although there were some improvements, forest condition in Europe has generally deteriorated.

The transnational survey results for 1994 revealed that 26.4 per cent of the total sample of around 102,300 trees were defoliated by more than 25 per cent and were thus classified as damaged. This is an increase of 3.8 percentage points compared to 1993 (22.6 per cent). The total tree samples of 1993 and 1994 can be compared, but any detailed analysis should be corrected for changes in the total sample.

In 1994, 12.1 per cent of the total tree sample showed signs of discoloration above 10 per cent. This value is 2.1 percentage points higher than in 1993.

As regards the two main species groups, 24.3 per cent of all broadleaves and 28.0 per cent of all conifers were found to be damaged in 1994, so the broadleaves are still in a slightly better condition. Among the most common species, the most severely affected broad-leaved species group was Quercus spp. (deciduous) with 32.4 per cent of trees damaged, followed by other broadleaves with 27.5 per cent of trees damaged. Abies spp. and Picea spp. were the most affected of the common coniferous species, with 32.9 per cent and 30.2 per cent, respectively, of trees damaged.

To trace the development of forest condition over several years without bias due to differences in the sam-

ples, special tree and plot samples which were common to certain survey years were analysed. Such common samples were determined for the periods 1993-1994 and 1988-1994.

The share of damaged common sample trees (CSTs) for 1993-1994 increased within that period from 23.1 per cent to 26.2 per cent. The largest increase (from 16.2 per cent to 21.9 per cent) occurred in the Mediterranean (higher) region, particularly in Quercus suber, Quercus ilex and Eucalyptus spp., and was mainly attributed to heat and drought. However, this last species showed the lowest damage in the Mediterranean region. A similarly high increase in damaged trees (from 27.2 per cent to 32.6 per cent) occurred in the Continental region, where Abies alba and Quercus spp. in Romania were particularly affected; damage was mainly attributed to drought and local air pollution. In the Sub-Atlantic region, the respective increase (from 39.3 per cent to 43.8 per cent) was partly due to summer drought, subsequent insect attack and local air pollution in Pinus sylvestris, Picea abies, Fagus sylvatica and Quercus spp. in the main damage areas of Germany, the Czech Republic, Poland and Slovakia.

In the subsample of common trees of the 1988 to 1994 surveys, the development of the defoliation of 12 species was analysed. With the exception of Abies alba, the proportion of damaged trees increased in all species during this period. Among the conifers, although Picea sitchensis still showed the greatest increase, from 2.3 per cent in 1988 to 20.3 per cent in 1994, there was some improvement from 1993 to 1994. The damage is thought to be mainly caused by attacks of Elatobium abietum (green spruce aphid). A similarly alarming deterioration appeared in Pinus halepensis, whose damaged share increased from 5.2 per cent in 1988 to 22.1 per cent in 1994. In contrast, the respective proportion of Abies alba diminished slightly from 25.8 per cent to 23.9 per cent; defoliation was at its worst in 1993 (30.1 per cent). Some improvement occurred in Quercus robur in 1994, but its damaged share increased markedly from 13.0 per cent in 1988 to 24.9 per cent in 1994.

In 1994, adverse weather conditions, particularly drought and heat, as well as insects, fungi, game, action of man, air pollution and forest fire were the most important probable causes of the observed defoliation and discoloration, as reported in both the national and transnational surveys. There were only a few reports of known pollution sources that had a direct impact on forest condition. However, there might be more widespread effects of air pollution which could not be found by this assessment.

According to the results of the national surveys, particularly the results from the main damage areas of some countries, but also those from several other regions, air pollution is considered as a major concern, because the atmospheric concentrations and the depositions of several air pollutants are thought to exceed the critical levels and loads for forest ecosystems. The countries where a high level of air pollution has been detected regard air pollution as the most important factor causing forest damage. The majority of the remaining countries consider air pollution as a predisposing, accompanying and locally triggering factor weakening forest ecosystems.

The survey results reveal a large spatial and temporal variation. It is therefore necessary to continue level I monitoring and to promote synoptic evaluations of its results together with other large-scale ecological parameters, in order to verify the effects caused by long-range transboundary air pollution. Plans are being made to evaluate the complete level I data set in the future.

In addition to level I monitoring, intensive monitoring (level II) has been carried out aimed at recognizing the factors and processes involved, especially the impact of air pollutants on the more common forest ecosystems. To this end, a soil inventory, foliar analyses, deposition measurements and increment studies are being conducted, on a number of selected permanent monitoring plots.

The report also outlines a possible future development in programme monitoring activities and in the assessment and reporting of the results.

INTRODUCTION

Forest condition in Europe has now been monitored on a large scale for nine years. The objective of this joint activity of the United Nations Economic Commission for Europe and the European Union is to acquire knowledge on the spatial and temporal variation of forest condition. To this end, an extensive monitoring approach (level I) is being used, based on a grid of systematically selected plots covering the forest area of participating countries (national grids of different densities) and of Europe (16 km x 16 km grid). Crown condition and a range of other ecological parameters are assessed on these plots. Moreover, on a great number of these plots soil and foliar analyses are also conducted.

In order to contribute to a better, more comprehensive understanding of cause-effect relationships, more intensive monitoring (level II) has also been carried out. This approach is based on a smaller number of monitoring plots situated in selected forest ecosystems, with a higher monitoring intensity per plot. Besides crown condition assessment and soil and foliage analyses, increment studies, deposition measurements and meteorological measurements are also carried out on level II.

The present report documents the results of the 1994 level I crown condition assessment, as carried out by the International Cooperative Programme on Assessment and Monitoring of Air Pollution Effects on Forests (ICP Forests) in cooperation with the European Union (EU). Based on the results of previous surveys, the development of crown condition since the beginning of monitoring is also described. Evaluations concerning the level I soil and foliar analyses, as well as the more intensive level II monitoring, will be the subject of future reports.

Chapter II describes the principles of the survey methods. Knowledge of these methods is essential for the interpretation of the results.

The results of the 1994 transnational and national surveys are presented in chapter III. The transnational results (section A) reflect forest condition in Europe without regard to national borders and refer to correlations between defoliation and discoloration and the site parameters. The national reports (section B) reflect the forest condition in individual countries with emphasis on its interpretation in connection with the multitude of damaging agents, particularly air pollution.

In chapter IV, both the transnational and the national survey results are interpreted together, also giving special attention to the effects of air pollution. These interpretations represent the views of the members of the two Programmes of ECE and EU.

Chapter V presents the conclusions drawn from the survey results and their interpretation.

The annexes provide tables relevant to the national results.

I. METHODS OF THE 1994 SURVEYS

A. Transnational survey

The transnational survey's objective is to document the spatial distribution and the development of forest condition on the European level. To reach this objective, the crown condition of forest trees is monitored on a large scale and a number of site parameters are assessed on a 16 km x 16 km transnational grid of sample plots. In several countries the plots of this transnational grid are a subsample of a denser national grid.

EU calculated the coordinates of the transnational grid and provided them to the participating countries. If a country had already established plots with different coordinates, the existing plots were accepted, provided that the mean point density resembled that of a 16 km x 16 km grid, and that the assessment methods corresponded to those of the ICP Forests Manual and the relevant European Commission Regulations. The fact that the grid is less dense in parts of the boreal forests is negligible considering the homogeneity and the current condition of these forests.

TABLE 1

Defoliation and discoloration classes according to ECE and EU classification

Defoliation class	Needle/leaf loss	Degree of defoliation
0	up to 10%	none
1	>10 - 25%	slight (warning stage)
2	>25 - 60%	moderate
3	>60%	severe
4	100%	dead
Discoloration class	Foliage discoloured	Degree of discoloration
0	up to 10%	none
1	>10 - 25%	slight
2	>25 - 60%	moderate
3	>60%	severe
4	100%	dead

B. National surveys

The objective of the national surveys is to document the forest condition and its development in the respective country. Therefore, the national surveys are conducted on national grids. The densities of these national grids vary between 1 km x 1 km and 32 km x 32 km due to differences in the size of forest area, in the structure of forests and in forest policies. Any comparisons between the national surveys of different countries should be made with great care because of differences in species composition, site conditions and reference trees.

C. Selection of sample trees

Ideally at least 20 sample trees are systematically selected according to a statistically sound procedure on each sampling point of the national and transnational grids, provided that it falls into forest land. Predominant, dominant, and co-dominant trees (according to the system of KRAFT) of all species qualify as sample trees, as long as they have a minimum height of 60 cm and that they do not show significant mechanical damage. Trees removed within management operations or blown over by wind or dead trees must be replaced by newly selected trees. A special evaluation of the data from 1988 to 1993 (Forest Condition Report 1994) showed that this replacement of trees does not distort the survey results.

D. Assessment parameters and data presentation

On each plot defoliation of the sample trees is assessed in comparison to a reference tree of full foliage; discoloration is also assessed. Photo guides suitable for the region under investigation may be used if no reference tree can be found in the vicinity of the sample trees.

In principle, the transnational survey results for defoliation are reported in 5 per cent bands, and the national survey results for defoliation according to the traditional classification (table 1). Most countries also report their national results for defoliation in 10 per cent bands. The assessment down to the nearest 5 or 10 per cent permits studies of the annual variation of foliage with far greater accuracy than the traditional system of only five classes of uneven width. Discoloration is reported according to the traditional classification both in the transnational and in the national surveys.

Changes in defoliation and discoloration attributable to air pollution cannot be differentiated from those caused by other factors. Consequently, defoliation due to other factors is included in the assessment results, although known causes should be recorded. However, major mechanical damage (for instance due to wind or snow) is ruled out as a cause as such trees are excluded from the sample.

In the presentation of the results a change is called "significant" if a statistical significance test was performed at a 95 per cent probability level.

Besides defoliation and discoloration, additional parameters have to be assessed on the plots of the transnational survey, as laid down in Commission Regulation (EEC) No. 1996/87. Within the transnational crown condition survey, the following plot and tree parameters have to be reported for each plot: country, plot number, plot coordinates, altitude, aspect, water availability, humus type, soil type (optional), mean age of dominant storey, tree numbers, tree species, observations of easily identifiable damage, date of observation.

The transnational survey results are submitted treewise and plotwise in digital format via EU or directly to the Programme Coordinating Centre (PCC) West of ICP Forests for screening, storage and evaluation. The national survey results are submitted on paper to the Coordinating Centre as country-related mean values, classified according to species and age groups. The data sets are accompanied by national reports providing explanations and interpretations.

TABLE 2

Percentages of defoliation for broadleaves, conifers, and all species

	Species type	Defoliation							Number of trees
		0-10%	>10-25%	0-25%	>25-60%	>60%	dead	>25%	
EU	Broadleaves	49.6	33.5	83.1	14.0	1.9	1.0	16.9	25280
	Conifers	47.2	34.3	81.5	15.7	1.4	1.5	18.6	23112
	All species	48.4	33.8	82.2	14.8	1.7	1.2	17.7	48392
Total Europe	Broadleaves	41.5	34.1	75.6	20.7	2.6	1.0	24.3	44449
	Conifers	37.6	34.4	72.0	24.8	2.1	1.1	28.0	57839
	All species	39.3	34.3	73.6	23.0	2.3	1.1	26.4	102288

TABLE 3

Percentages of discoloration for broadleaves, conifers, and all species

	Species type	Discoloration					Number of trees
		0-10%	>10-25%	>25-60%	>60%	dead	>10%	
EU	Broadleaves	87.9	8.3	2.0	0.8	1.0	12.1	25280
	Conifers	86.8	9.6	1.6	0.4	1.5	13.1	23108
	All species	87.4	8.9	1.8	0.6	1.2	12.5	48388
Total Europe	Broadleaves	87.1	8.7	2.4	0.7	1.1	12.9	43537
	Conifers	88.5	8.2	1.8	0.3	1.2	11.5	53541
	All species	87.9	8.4	2.1	0.5	1.1	12.1	97078

The results of the evaluation are presented mainly in terms of the percentages of the tree sample falling into the traditional five defoliation or discoloration classes. This classification reflects to a certain extent the experience gathered in forest damage assessments in central Europe between 1980 and 1983. At that time, any loss of foliage exceeding 10 per cent was considered as abnormal, indicating an incipient stage of impaired forest health. Assumptions based on physiological investigations of the vitality of differently defoliated trees led to the establishment of uneven class widths. Because of these reasons and in order to ensure comparability with previous presentations of survey results, the traditional classification of both defoliation and discoloration has been retained for comparative purposes, although it is considered arbitrary by some countries.

A certain natural range is taken into account by defining a level of defoliation up to 25 per cent as "undamaged". A defoliation of > 10-25 per cent indicates a "warning-stage". Therefore, in the present report a distinction has often been made only between defoliation classes 0 and 1 (0-25 per cent defoliation) on the one hand, and classes 2, 3 and 4 (defoliation > 25 per cent) on the other.

Classes 2, 3 and 4 represent trees with considerable defoliation and are thus referred to as "damaged". Like

the sample trees, the sample points are referred to as "damaged" if the mean defoliation of their trees (expressed as percentages) falls into class 2 or higher. Otherwise the sample point is considered as "undamaged".

The most important results have been tabulated separately for all participating countries (called "total Europe") and for those 12 countries that were EU member States in the 1994 survey year.

II. RESULTS OF THE 1994 SURVEYS

A. Transnational survey results

1. General results

The 1994 transnational survey was carried out in 29 countries, comprising all 12 EU member States and 17 non-EU countries. With two non-EU countries more than in 1993 (Bulgaria and Latvia), the number of countries participating in the survey was the largest ever.

Also, with 102,288 trees assessed on 4,756 plots, the database is now larger than ever before. Compared to 26,084 trees assessed in the first survey in 1987, the

TABLE 4

Percentages of defoliation of all species by mean age

| | Mean age | Defoliation | | | | | | | Number |
	[years]	0-10%	>10-25%	0-25%	>25-60%	>60%	dead	>25%	of trees
EU	0 - 20	63.6	24.7	88.3	8.0	1.5	2.2	11.7	7654
	21 - 40	54.2	30.0	84.2	12.7	2.0	1.1	15.8	12361
	41 - 60	46.6	35.5	82.1	14.9	1.6	1.4	17.9	8276
	61 - 80	39.6	41.8	81.4	16.3	1.7	0.6	18.6	5444
	81 -100	41.3	39.3	80.6	17.9	0.8	0.7	19.4	4816
	101-120	32.7	40.8	73.5	23.6	2.5	0.4	26.5	2428
	>120	31.4	37.4	68.8	29.1	1.8	0.3	31.2	2701
	Irregular	47.2	35.5	82.7	13.9	1.6	1.8	17.3	4712
	Total	48.5	33.8	82.3	14.8	1.7	1.2	17.7	48392
Total	0 - 20	61.8	24.4	86.2	9.7	2.1	2.0	13.8	9154
Europe	21 - 40	52.6	29.4	82.0	14.8	2.1	1.1	18.0	20242
	41 - 60	37.3	36.0	73.3	23.1	2.4	1.2	26.7	20175
	61 - 80	30.7	39.4	70.1	27.0	2.2	0.7	29.9	17269
	81 -100	33.5	37.0	70.5	27.0	1.9	0.6	29.5	12693
	101-120	30.6	38.4	69.0	27.8	2.9	0.3	31.0	5690
	>120	32.0	36.0	68.0	28.4	3.1	0.5	32.0	5770
	Irregular	46.5	35.2	81.7	15.1	1.8	1.4	18.3	6023
	Total	41.0	34.4	75.4	21.4	2.2	1.0	24.6	97016

database is now approximately four times as large as at the beginning of the programme. This extension is partly due to the completion of the grid within EU member States, but mainly the consequence of the participation of a growing number of non-EU countries since 1990.

Defoliation was assessed on all sample trees in total Europe and on 48,392 trees in the EU member States. In total Europe, the share of sample trees considered as damaged, i.e. with defoliation above 25 per cent, amounted to 26.4 per cent. In the EU member States the respective share was 17.7 per cent. The conifers had a higher proportion of damaged trees (28.0 per cent) than the broadleaves (24.3 per cent) in total Europe. As in previous years, this difference was less pronounced in the EU member States (18.6 per cent and 16.9 per cent, respectively). Table 2 shows the results in greater detail.

As several non-EU member States did not assess discoloration on all their sample trees, discoloration was reported for only 97,078 trees in 1994; 12.1 per cent of this tree sample had a discoloration of more than 10 per cent (table 3).

An evaluation of the spatial distribution of the percentages of damaged trees per plot over the entire survey area reveals that on 48.1 per cent of the plots the share of damaged trees is 10 per cent or less. These plots are mainly located in south-western Europe, Scandinavia and in the eastern part of the Alps. On the other hand, the share of damaged trees ranges from 51 per cent to 75 per cent on 10.1 per cent of the plots, and from 76 per cent to 100 per cent on 10.3 per cent of the plots. This means that on 20.4 per cent of all plots more than half the trees are damaged. As in previous years, the areas with the highest proportion of damaged trees are mainly located in central Europe.

On 30.8 per cent of the plots the mean defoliation (classified according to the five defoliation classes) is higher than 25 per cent (classes 2-4 with 29.8 per cent, 0.8 per cent and 0.2 per cent, respectively). Such a situation is particularly frequent in central Europe.

2. Forest condition by species groups

Among the broadleaves of the total tree sample, defoliation was the highest for *Quercus* spp. (32.4 per cent damaged). The lowest percentage of damaged trees was found for *Quercus suber* with 14.2 per cent and *Eucalyptus* spp. with 10.6 per cent. Of the conifers, *Abies* spp. had the highest percentage of damaged trees (32.9 per cent). The lowest share of damaged trees was recorded for *Larix* spp. (19.1 per cent).

Discoloration among the broadleaves of the total tree sample was most obvious for *Castanea sativa* (20.0 per cent of the trees discoloured, i.e. showing discoloration greater than 10 per cent). *Betula* spp. had the lowest percentage of discoloured trees (4.1 per cent). Among the conifers the interspecific variation was smaller. In total Europe *Abies* spp. was the species group with the highest percentage of discoloured trees (21.7 per cent). The least discoloration was found in *Larix* spp., with 7.8 per cent of the trees discoloured.

3. Defoliation and discoloration by mean age

For both the EU member States and total Europe, tables 4 and 5 show the percentages of trees in each defoliation and discoloration class for seven classes of different mean stand age and for a class of irregular age composition.

TABLE 5

Percentages of discoloration of all species by mean age

	Mean age [years]	Discoloration						Number of trees
		0-10%	>10-25%	>25-60%	>60%	dead	>10%	
EU	0 - 20	84.6	10.1	2.4	0.7	2.2	15.4	7651
	21 - 40	85.7	10.6	1.8	0.8	1.1	14.3	12360
	41 - 60	90.3	6.8	1.2	0.3	1.4	9.7	8276
	61 - 80	89.5	6.9	1.9	1.1	0.6	10.5	5444
	81 -100	90.1	7.9	1.1	0.1	0.8	9.9	4816
	101-120	89.4	7.5	1.7	1.0	0.4	10.6	2428
	>120	90.6	6.9	2.1	0.1	0.3	9.4	2701
	Irregular	82.7	12.0	2.7	0.8	1.8	17.3	4712
	Total	87.5	8.9	1.8	0.6	1.2	12.5	48388
Total Europe	0 - 20	84.4	10.3	2.7	0.6	2.0	15.6	9150
	21 - 40	85.3	10.3	2.6	0.7	1.1	14.7	20239
	41 - 60	89.3	7.0	2.2	0.3	1.2	10.7	20165
	61 - 80	91.0	6.4	1.4	0.5	0.7	9.0	17266
	81 -100	89.2	8.3	1.7	0.2	0.6	10.8	12680
	101-120	89.6	7.6	1.8	0.7	0.3	10.4	5689
	>120	89.8	7.9	1.6	0.2	0.5	10.2	5766
	Irregular	83.7	11.7	2.5	0.7	1.4	16.3	6023
	Total	88.0	8.4	2.1	0.5	1.0	12.0	96978

As in the previous years, the 1994 survey gives evidence of the strong positive correlation between age and defoliation. The share of damaged trees gradually increases with increasing mean age between ages 0-80. At higher ages, however, the share of damaged trees remains approximately the same.

The shares of trees in different discoloration classes do not vary greatly with age. The younger trees (0-40 years) and the older trees (81-120 years) seem to be slightly more discoloured than the trees between 41 and 80 years of age.

4. Easily identifiable damage

The eight easily identifiable types of damage are:

—Game and grazing (damage to trunk, bark, etc.);

—Presence or traces of an excessive number of insects;

—Fungi;

—Abiotic agents (wind, drought, snow, etc.);

—Direct action of man (poor silvicultural practices, logging, etc.);

—Fire;

—Known local or regional pollution (classic smoke damage);

—Other types of damage.

For these categories, only the presence of such damage is indicated. It is presented in table 6 in terms of the percentage of the total tree or plot sample. No indication is given of the intensity of the damage. It is possible that

more than one type of identifiable damage occurs on a single tree. Such trees will therefore be represented more than once in the table. Of the 102,288 trees of the total tree sample, 23,083 trees (22.6 per cent) were reported to have identifiable damage due to one or more causes. These trees were found on 2,722 plots (57.2 per cent) of the total plot sample. On the other trees identifiable damage was either not present or not assessed.

As in the previous years, the most commonly observed type of damage in total Europe was caused by insects (9.3 per cent of the trees and 23.4 per cent of the plots). The second and third most commonly observed types were abiotic agents and fungi affecting 4.9 per cent and 4.6 per cent of the total tree sample.

Damage due to the action of man and other damage were observed less frequently, representing 3.4 per cent and 3.2 per cent, respectively, of the total tree sample.

Game/grazing, fire and damage by known pollution (i.e. classic smoke damage caused by air pollution from nearby emitters) were observed to a far lesser extent, namely on 1.2 per cent, 0.5 per cent and 0.3 per cent of the trees, respectively. Of the total sample, 4.1 per cent of the trees suffered damage from more than one cause.

Among the trees showing any identifiable damage, the proportions of trees in defoliation classes 2-4 ranged from 0.1 per cent (103 trees) (known pollution) to 3.1 per cent (3,131 trees) (insects) in total Europe.

Of all trees showing any identifiable damage, the percentage of trees in defoliation classes 2-4 was 7.0 per cent. This figure together with the share of trees showing no identifiable damage (19.4 per cent) totals 26.4 per cent, which is the share of trees in defoliation classes

TABLE 6

Percentages of trees with defoliation > 25 per cent and discoloration > 10 per cent by identified damage types, based on a total of 4,756 plots with 102,288 (defoliation) and 97,078 (discoloration) trees, respectively

Damage type	Defoliation % in classes 2, 3, 4 (>25%) of total trees		Discoloration % in classes 1, 2, 3, 4 (>10%) of total trees		Observations			
					% of total trees		% of total plots	
	Total Europe	EU	Total Europe	EU	Total Europe	EU	Total Europe	EU
Game/Grazing	0.3	0.3	0.1	0.2	1.2	1.1	4.4	2.6
Insects	3.1	2.3	1.7	1.6	9.3	9.9	23.4	28.3
Fungi	1.5	0.9	1.2	1.2	4.6	3.4	18.5	14.6
Abiotic agents	1.8	0.9	1.4	1.3	4.9	2.9	21.1	12.8
Action of man	0.9	0.8	0.6	0.8	3.4	3.1	13.7	7.2
Fire	0.4	0.7	0.3	0.6	0.5	0.9	0.7	1.3
Known pollution	0.1	0.0	0.2	0.1	0.3	0.1	0.4	0.1
Other	0.8	0.6	0.5	0.5	3.2	2.0	18.2	8.1
Any ident. damage	7.0	5.5	4.6	4.9	22.6	20.2	57.2	49.5
No ident. damage (or not assessed)	19.4	12.2	7.5	7.7	77.4	79.8	42.8	50.5
Total	26.4	17.7	12.1	12.6	100.0	100.0	100.0	100.0

2-4 of the total tree sample. It must be noted, however, that the share of trees showing no identifiable damage comprises an unknown amount of trees not assessed for easily identifiable damage. The same applies to discoloration.

The most pronounced negative effect in terms of discoloration was also observed on trees affected by insects—1.7 per cent of all trees.

Data on identifiable damage permit only very general conclusions on common and widespread damage to be drawn and do not lead to any conclusions on cause-effect relationships. Defoliation and discoloration are suited only to describe the general tree condition. Many stress factors could be identified only after more detailed studies as done by the level II forest monitoring.

5. Changes in defoliation and discoloration from 1993 to 1994

The total tree samples of 1993 and 1994 can be compared, but detailed analysis should be corrected for changes in the total sample (e.g. because of an increasing number of participating countries).

For a comparison of the 1993 and 1994 survey results, a subsample called common sample trees (CSTs) is defined containing all trees that are common to both surveys. For 1993 and 1994, this common sample consisted of 86,085 trees, representing 83.7 per cent of the total tree sample of 1993 and 84.2 per cent of the total tree sample of 1994. This is 1,116 trees or 1.3 per cent more than in the 1993 survey. The reason for this slight increase in the number of CSTs is the participation of Croatia, Estonia, the Republic of Moldova and Slovenia in the transnational forest condition assessment since 1993.

The common sample of 1993 and 1994 was the largest ever. The increasing number of CSTs improves the reliability of the calculation of changes in defoliation and discoloration and the consistency between the data sets in the participating countries.

The percentages of trees in the different defoliation and discoloration classes for the total tree samples and for the CSTs of 1993 and 1994 are shown in table 7.

6. Changes by climatic region

As in previous years, the total tree sample and common sample trees (CSTs) were classified into climatic regions in order to account for various climatic site conditions. The selected climatic regions largely match the most important forest vegetation types. In addition to the nine climatic regions specified in the 1993 and 1994 reports, a new "Mountainous (north)" region is separated from the "Mountainous" region in order to account for the climatic differences between northern and southern Europe.

The percentages of damaged trees and mean plot defoliation were used to quantify the changes in defoliation of the CSTs from 1993 to 1994 for each climatic region. Figure 1 visualizes the changes in the percentage of trees in the defoliation classes. The following descriptions refer to the changes in the percentage of trees damaged and differences in mean defoliation between 1993 and 1994.

In terms of differences in mean defoliation, significant changes were found for the total CSTs of all regions and for each climatic region as well. Except for the Boreal and Atlantic (north) regions, the mean defoliation increased significantly between 1993 and 1994, but in no case did the change reach the 5 per cent mark, which is the actual assessment accuracy in the field. The most pronounced deterioration in crown condition in terms of

TABLE 7

Percentages of the total tree sample and the common sample trees in different defoliation and discoloration classes in 1993 and 1994

	Total tree sample		Common Sample Trees	
	1993	1994	1993	1994
Defoliation				
0-10%	43.5	39.3	42.9	39.6
>10-25%	33.9	34.3	34.0	34.2
0-25%	77.4	73.6	76.9	73.8
>25-60%	19.9	23.0	20.8	22.8
>60%	1.9	2.3	2.0	2.3
dead	0.8	1.1	0.3	1.1
>25%	22.6	26.4	23.1	26.2
Number of trees	102 800	102 288	86 085	86 085
Discoloration				
0-10%	90.0	87.9	89.9	88.7
>10-25%	7.2	8.4	7.7	8.0
>25-60%	1.7	2.1	1.8	1.7
>60%	0.3	0.5	0.3	0.5
dead	0.8	1.1	0.3	1.1
>10%	10.0	12.1	10.1	11.3
Number of trees	86 461	97 078	82 005	82 005

the percentage of damaged trees occurred in the Mediterranean (higher) region (5.7 percentage points) and in the Continental region (4.5 percentage points). The increase in the percentages of damaged trees in the Mountainous (north) and the Mountainous (south) region was less obvious, however the shifts are statistically significant. The improvements in crown condition in the Boreal and Atlantic (north) regions proved to be significant only in terms of differences in mean defoliation between 1993 and 1994.

The changes in the percentages of discoloured trees in each climatic region are documented in figure 2. Except for the Atlantic (south) region and the Mountainous regions the results of changes in discoloration correspond to those of defoliation, revealing a slightly deteriorated health status in 1994 as compared to 1993. The percentage of discoloured trees rose markedly in the Mediterranean (lower) region (by 6.4 percentage points), followed by the Continental region with an increase of 5.7 percentage points. Only in the Atlantic (south) region did discoloration decrease significantly (by 5.0 percentage points). Also decreasing, but not significantly, were the percentages of discoloured trees in the Mountainous (north) and Mountainous (south) regions (-1.3 and -0.2 percentage points, respectively).

7. Changes by species group

In 1994, the CSTs as a whole showed a significant deterioration in defoliation. The share of damaged CSTs increased from 23.1 per cent in 1993 to 26.1 per cent in 1994. In the broad-leaved CSTs the proportion of trees with a defoliation above 25 per cent rose from 20.0 per cent to 23.6 per cent. In the coniferous CSTs the respective proportion also increased, namely from 25.4 per cent to 28.2 per cent.

Some of the species among the broad-leaved CSTs showed a remarkable deterioration, as expressed by the shares of damaged trees. The crown condition of *Eucalyptus* spp., *Quercus ilex* and *Quercus suber* deteriorated notably. The proportion of damaged *Eucalyptus* spp. increased from 4.1 per cent to 11.5 per cent. However, this species showed the lowest damage figures in the Mediterranean area. The share of damaged *Quercus ilex* trees rose from 6.9 per cent to 14.8 per cent. The respective proportion of *Quercus suber* increased from 9.0 per cent to 14.4 per cent. A decrease in defoliation occurred only among *Carpinus* spp., the damaged share of which diminished from 25.9 per cent to 22.1 per cent.

As in the previous report, the rapid changes in vitality among the principal Mediterranean species *Eucalyptus* spp., *Quercus ilex* and *Quercus suber* should be interpreted in connection with typical detrimental events in the Mediterranean region, such as drought and fire, especially if only small percentages of trees are affected. Though large, these changes have less influence on the result for the total broadleaves, due to the low numbers of CSTs among these species groups.

The deciduous *Quercus* spp. with 10,798 trees represented the largest number of broad-leaved CSTs. Consequently, their increase in the proportion of damaged trees from 26.8 per cent to 30.1 per cent greatly influenced the result for the broad-leaved CSTs. Also influential were the "other broadleaves" with 6,975 trees and an increase in damaged trees from 20.2 per cent to 26.7 per cent.

Most species groups of the coniferous CSTs experienced only slight changes in defoliation between 1993 and 1994, except "other conifers", whose share of damaged trees increased considerably, from 14.5 per cent to 21.3 per cent, and *Abies* spp., which showed a decrease from 34.0 per cent to 31.9 per cent. Nevertheless, *Abies* spp. had the highest percentage of damaged trees in 1993

FIGURE 1

Percentages of defoliation of the common sample trees in 1993 and 1994 for each of the 10 climatic regions and for the total sample of CSTs

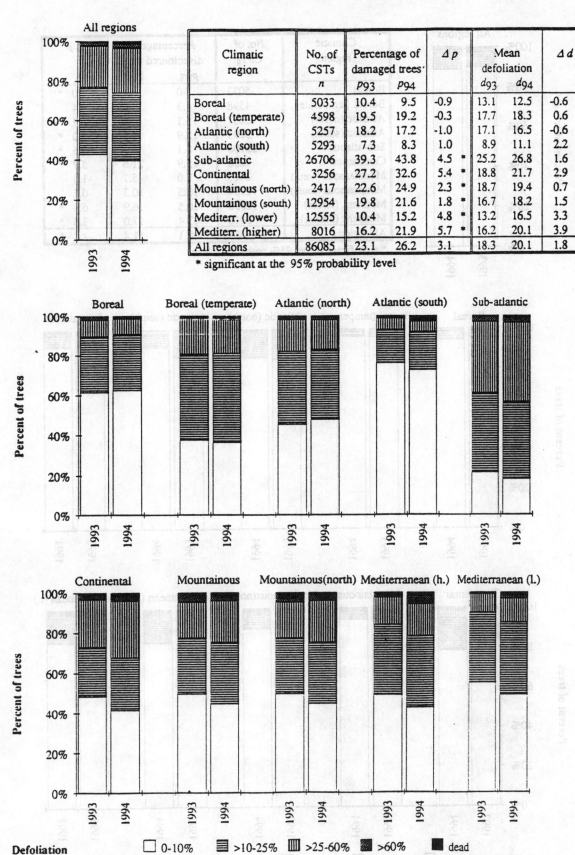

Climatic region	No. of CSTs n	Percentage of damaged trees $P93$ $P94$		Δp	Mean defoliation d_{93} d_{94}		Δd	
Boreal	5033	10.4	9.5	-0.9	13.1	12.5	-0.6	*
Boreal (temperate)	4598	19.5	19.2	-0.3	17.7	18.3	0.6	*
Atlantic (north)	5257	18.2	17.2	-1.0	17.1	16.5	-0.6	*
Atlantic (south)	5293	7.3	8.3	1.0	8.9	11.1	2.2	*
Sub-atlantic	26706	39.3	43.8	4.5 *	25.2	26.8	1.6	*
Continental	3256	27.2	32.6	5.4 *	18.8	21.7	2.9	*
Mountainous (north)	2417	22.6	24.9	2.3 *	18.7	19.4	0.7	*
Mountainous (south)	12954	19.8	21.6	1.8 *	16.7	18.2	1.5	*
Mediterr. (lower)	12555	10.4	15.2	4.8 *	13.2	16.5	3.3	*
Mediterr. (higher)	8016	16.2	21.9	5.7 *	16.2	20.1	3.9	*
All regions	86085	23.1	26.2	3.1	18.3	20.1	1.8	*

* significant at the 95% probability level

Defoliation □ 0-10% ▤ >10-25% ▥ >25-60% ▧ >60% ■ dead

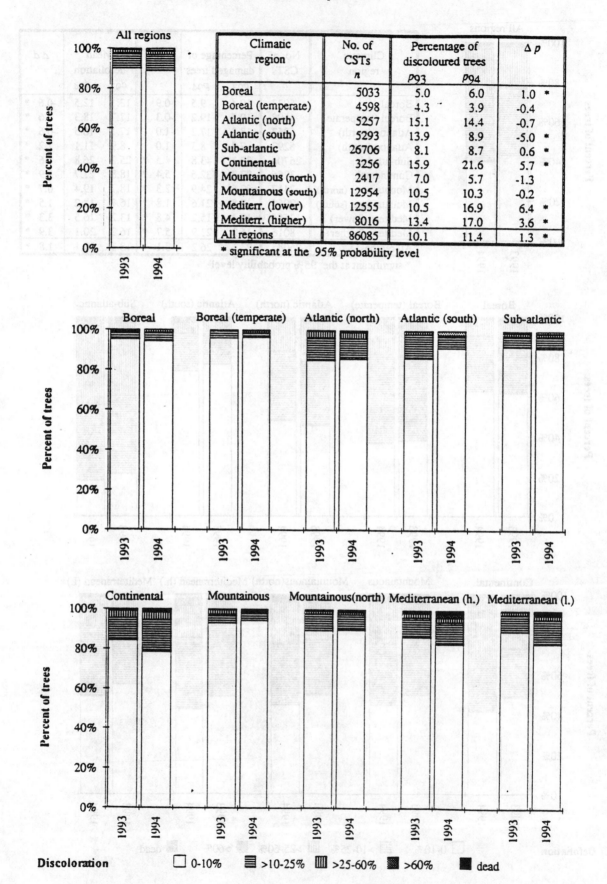

FIGURE 2

Percentages of discoloration of the common sample trees in 1993 and 1994 for each of the 10 climatic regions and for the total sample of CSTs

Climatic region	No. of CSTs n	Percentage of discoloured trees		Δ p	
		p93	p94		
Boreal	5033	5.0	6.0	1.0	*
Boreal (temperate)	4598	4.3	3.9	-0.4	
Atlantic (north)	5257	15.1	14.4	-0.7	
Atlantic (south)	5293	13.9	8.9	-5.0	*
Sub-atlantic	26706	8.1	8.7	0.6	*
Continental	3256	15.9	21.6	5.7	*
Mountainous (north)	2417	7.0	5.7	-1.3	
Mountainous (south)	12954	10.5	10.3	-0.2	
Mediterr. (lower)	12555	10.5	16.9	6.4	*
Mediterr. (higher)	8016	13.4	17.0	3.6	*
All regions	86085	10.1	11.4	1.3	*

* significant at the 95% probability level

Discoloration   ☐ 0-10%   ≡ >10-25%   ▥ >25-60%   ▦ >60%   ■ dead

FIGURE 3

FIGURE 3

Development of defoliation for coniferous trees (defoliation classes 2-4) common to 1988-1994

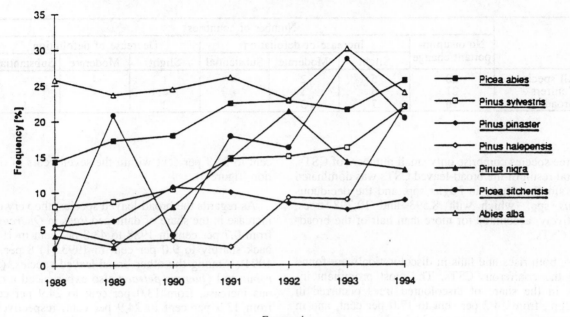

FIGURE 4

Development of defoliation for broad-leaved trees (defoliation classes 2-4) common to 1988-1994

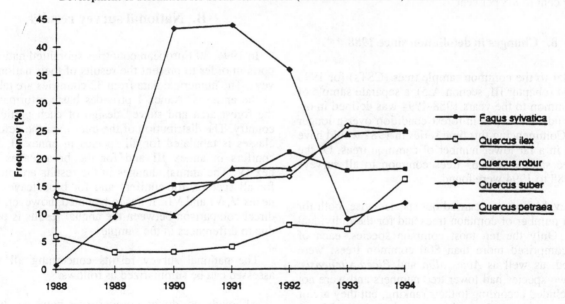

and 1994, both among the conifers and the broadleaves. However, with 1,979 trees, *Abies* spp. had only little influence on the total coniferous result, which is dominated mainly by *Pinus* spp. with 27,142 trees and *Picea* spp. with 17,258 trees.

The proportion of damaged *Pinus* spp. trees rose from 23.7 per cent to 27.0 per cent. The share of damaged *Picea* spp. trees increased from 28.3 per cent to 30.8 per cent. The proportion of damaged coniferous CSTs increased from 25.4 per cent to 28.2 per cent, mainly as a result of the deterioration in these most comprehensive species groups.

Overall, discoloration was worse in 1994 than in 1993, both in the conifers and in the broadleaves. As in the previous year, some species groups improved over the 1993-1994 period, whereas other species groups deteriorated.

Among the broad-leaved CSTs, the share of discoloured *Eucalyptus* spp. (discoloration classes 1-4) increased considerably from 5.5 per cent to 17.4 per cent. In contrast, the respective proportion of *Carpinus* spp. fell sharply, from 26.3 per cent to 12.9 per cent. Further obvious increases in discoloration occurred in *Quercus ilex* (from 4.5 per cent to 7.8 per cent) and in "other broadleaves" (from 12.9 per cent to 14.6 per cent). Other notable drops in discoloration were found in *Castanea sativa* (from 23.6 per cent to 20.8 per cent) and in *Betula* spp. (from 5.1 per cent to 3.7 per cent). However, with the exception of "other broadleaves",

TABLE 8

Changes in defoliation observed between 1993 and 1994 in classes 2-4

	No or unim-portant change	Increase of defoliation			Decrease of defoliation		
		Slight	Moderate	Substantial	Slight	Moderate	Substantial
All species	20	5	2	-	1	-	-
Conifers	21	5	1	-	4	-	-
Broadleaves	16	7	3	-	2	-	1

these tree species comprise only small numbers of CSTs. The total result of the broad-leaved CSTs was dominated by the small changes in *Fagus* spp. and the deciduous *Quercus* spp., which with 8,593 and 10,798 trees, respectively, accounted for more than half of the broad-leaved CSTs.

Also, both rises and falls in discoloration were found among the coniferous CSTs. The most prominent increases in the share of discoloured trees occurred in *Abies* spp., from 14.2 per cent to 17.0 per cent, and in *Pinus* spp., from 7.5 per cent to 11.1 per cent. A notable decrease was recorded for "other conifers", from 12.8 per cent to 8.8 per cent.

8. Changes in defoliation since 1988

Similar to the common sample trees (CSTs) for 1993 and 1994 (chapter III, section A.5), a separate sample of trees common to the years 1988-1994 was defined in order to study the trends in forest condition over a longer period. Commencing this time series in 1987 would have resulted in a far lower number of common trees. Of the total tree sample, 28,263 trees common to all surveys from 1988 to 1994 were found.

The evaluation was carried out species-wise, both for the total number of common trees and for the individual regions. Only the ten most common species, each of which comprised more than 800 common trees, were evaluated, as well as *Abies alba* and *Picea sitchensis*. These two species had lower tree numbers and were not to be included according to their ranking, but they are of importance in particular regions, especially in the Mountainous and in the Atlantic (north) regions. As in the previous surveys, there was no evaluation for those regions in which the number of trees of a certain species was lower than 100. No common trees existed in the Boreal, the Boreal (temperate) and the Continental regions.

Between 1988 and 1994 the proportions of the trees classified as damaged in the subsample of common trees differed considerably depending on the individual tree species. The 12 analysed species showed a more or less obvious increase in the proportion of damaged trees.

Among the conifers, *Picea abies* had the highest percentage of damaged trees in 1994 (25.6 per cent), but *Picea sitchensis* (2.3 per cent to 20.3 per cent) and *Pinus halepensis* (5.2 per cent to 22.1 per cent) experienced the most obvious increase since 1988. The proportion of damaged *Pinus sylvestris* trees rose steadily from 8.1 per

cent to 21.9 per cent within the seven years of observation (figure 3).

As regards the broad-leaved species, the very obvious increase in the share of damaged trees in *Quercus suber*, from 0.7 per cent in 1988 to 43.8 per cent in 1991, fell back sharply to 9.0 per cent in 1993 (11.8 per cent in 1994). Among the other broad-leaved species, *Quercus robur* and *Quercus petraea* also experienced a continuous increase, from 13.0 per cent to 24.9 per cent and from 13.9 per cent to 24.9 per cent, respectively (figure 4).

B. National survey results

In 1994, 30 European countries submitted national reports in order to present the results of their national surveys. The numerical data from 32 countries are tabulated in the annexes. Annex I provides basic information on the forest area and survey design of each participating country. The distribution of the trees over the defoliation classes is tabulated for all species in annex II, for the conifers in annex III and for the broadleaves in annex IV. The annual changes in the results are presented for all species, for conifers and for broadleaves in annexes V, VI and VII. It has to be noted, however, that no direct comparison between the annual results is possible due to differences in the samples.

The national survey results concerning all species assessed can be summarized as follows.

Although no direct comparisons between different countries are possible because of differences in the application of the common methodology and general variations in climatic and site factors, the data suggest that there are three groups of countries.

As in the previous year, in Ireland, the Russian Federation and Sweden only conifers were assessed. In four countries, namely Austria, France, Portugal and the Russian Federation, the percentage of sample trees classified as damaged (defoliation classes 2-4) was lower than 10 per cent.

In ten of the countries the percentage of sample trees classified as damaged ranged from 10 per cent to 20 per cent. These countries are Belgium (including Flanders and Wallonia), Estonia, Finland, Ireland, Italy, the Netherlands, Slovenia, Spain, Sweden, and the United Kingdom.

In another 18 countries, namely Belarus, Bulgaria, Croatia, the Czech Republic, Denmark, Germany, Greece, Hungary, Latvia, Lithuania, Luxembourg, the Republic of Moldova, Norway, Poland, Romania, Slovakia, Switzerland and Ukraine, the percentage of sample trees classified as damaged was greater than 20 per cent, with a maximum of 59.7 per cent. They represent more than half the member States which reported survey results. In ten of these countries the defoliation was particularly high in coniferous stands. The broad-leaved stands were particularly affected in Bulgaria, Germany, Greece, Luxembourg, Norway and Romania.

A deterioration has occurred in 21 countries which reported survey results. Table 8 describes the changes in defoliation observed between 1993 and 1994 in classes 2-4 by referring to all 32 countries which submitted survey results. Changes are rated as unimportant if equal to or less than 5.0 percentage points, as slight between 5.1 and 10.0 percentage points, as moderate between 10.1 and 20.0 percentage points and as substantial if exceeding 20.0 percentage points from one year to the next.

As regards all species, a slight increase in defoliation occurred in five countries, whereas a slight decrease was observed in only one country. Changes in defoliation are obvious in both the conifers and the broadleaves. Concerning the conifers, a rise occurred in six countries, whereas a fall was observed in four. In one country the increase in the conifers was moderate, but no substantial increase was found anywhere. In comparison to 1993, defoliation among the broadleaves clearly worsened. In seven countries a slight and in three countries a moderate increase occurred. However, in none of the countries was there a substantial increase in the broadleaves. Besides a slight decrease in two countries, there was a substantial decrease in one country.

III. INTERPRETATION

In 1994, a total of 32 European countries submitted their national forest inventory data. Twenty-nine of these countries also submitted results from the 16 km x 16 km grid (transnational survey). This was the largest number of countries ever participating in the transnational survey since its establishment in 1987. Of the 102,288 trees assessed on 4,756 plots, 26.4 per cent were considered as damaged. Valuable information provided in the national reports was often utilized for the interpretation of the survey results.

As in the previous years, the areas of highest defoliation are located in central Europe, but defoliation is also high in certain areas of northern and south-eastern Europe.

While differences in defoliation between different climatic regions cannot be readily attributed to climatic conditions, the national reports of two thirds of the countries explain their forest condition in terms of weather conditions. Weather conditions in 1994 were mentioned mainly as a cause of the deterioration, but partly also of the recuperation of forest condition. Nearly half the na-

tional reports emphasized that summer heat and drought in 1994 and in previous years had acted as a predisposing or triggering factor for the damage observed. Pests are often looked upon as secondary agents, which were fostered by hot and dry weather conditions. An improvement in forest condition was often attributed to high winter and spring precipitation, which compensated for the summer drought that had prevailed in many parts of Europe. These explanations from the national reports generally accord with the defoliation trends as derived from the transnational results. The changes in mean plot defoliation from 1993 to 1994 indicate a significant worsening in forest condition in central, eastern, south-eastern and southern Europe. Many countries in these regions also reported a deterioration in forest condition due to drought and heat. The regions in central and western Europe, for which a significant improvement in forests can be derived from the transnational results, partly coincide with the countries that reported an improvement in forest condition due to higher precipitation. In northern Europe, the harsh winter climate is mentioned as a typical stressor.

As in all previous surveys, the parameter showing the closest correlation with defoliation was stand age. The transnational survey revealed an increase in the percentage of damaged trees from 13.8 per cent in age class 0-20 years to 29.9 per cent in age class 61-80 years. With higher ages, the percentage of damaged trees increased slightly to 32.0 per cent in age class >120 years. This is generally consistent with the national survey results presented in previous reports, which have shown trees over 60 years old to be clearly more defoliated than younger ones. The close correlation between defoliation and stand age partly reflects the well-known natural loss of foliage due to ageing, and partly the fact that old trees are in general more susceptible to various environmental stress factors.

Easily identifiable damage was reported for 22.6 per cent of all trees of the transnational survey. As in recent years, the most frequently reported damage type was insect attack, affecting 9.3 per cent of the total tree sample. Second on the frequency scale were abiotic agents (4.9 per cent), followed by fungi (4.6 per cent) and action of man (3.4 per cent). This illustrates the multitude of stressors responsible for the defoliation assessed. Classic smoke damage was reported for a very small percentage of the total tree sample (0.3 per cent). For 77.4 per cent of the trees no evident source of damage was reported. These trees, however, comprise an unknown proportion of trees on which damage was present, but not reported.

The percentage of damaged trees in 1994 (26.4 per cent) was clearly higher than that in 1993 (22.6 per cent) or any other survey year. However, as mentioned above, direct comparisons between the shares of damaged trees in the total sample are biased because the samples differ from year to year. As a consequence, differences between the 1994 results and the results of previous years observed within the total sample or the subsamples of individual climatic regions do not give evidence of the actual development of forest condition. The actual trend in forest condition in Europe is better reflected by the common sample trees (CSTs) evaluated for different periods (1993-1994, 1990-1994 and 1988-1994).

With 86,085 CSTs for 1993 and 1994, this common sample was larger than ever before, which improves the consistency of the total database. The statistical evaluation of the 86,085 CSTs shows that the share of damaged trees rose from 23.1 per cent in 1993 to 26.2 per cent in 1994. Although this change was not statistically significant for the total CSTs, statistically significant changes were found in several climatic regions. These changes at the regional scale were often consistent with the reports received from the respective countries, as shown in the following paragraphs.

The largest increase occurred in the Mediterranean (higher) region, where the share of damaged trees increased significantly, up 5.7 percentage points, from 16.2 per cent in 1993 to 21.9 per cent in 1994. The change was also high in the Mediterranean (lower) region, with a significant increase, up 4.8 percentage points, from 10.4 per cent to 15.2 per cent. With respect to the two Mediterranean regions, the deterioration in forest condition is widespread in Spain, mainly in the northern and eastern parts, in Italy, as well as in certain parts of Greece. The species most affected were *Eucalyptus* spp., *Quercus ilex* and *Quercus suber*. This is consistent with the findings documented in the national reports from these countries. The national reports point to heat and drought as the main causes of the deterioration.

Another significant increase in the share of damaged CSTs was found in the Continental region, namely from 27.2 per cent to 32.6 per cent, i.e. up 5.4 percentage points. As in last year's survey, this increase reflects mainly the deterioration in forest condition in Romania, where continuing drought and local air pollution caused an increase in defoliation in many species, particularly in *Abies alba* and *Quercus* spp.

The results of the transnational and national surveys give evidence of a large variety of stressors and site conditions influencing the extent of defoliation and discoloration. As the most important causes, the countries have reported adverse weather conditions, insects, fungi, forest fires, game, action of man and air pollution. The degree to which individual factors have contributed to the defoliation and discoloration cannot be quantified, however, as these symptoms are non-specific. With the exception of a small (0.3 per cent) proportion of trees, damage by air pollution has not been unequivocally identified from the survey data. However, on the basis of more symptom-specific surveys of air pollution injury and/or a range of experimental research, more than half of the participating countries mention air pollution in their national reports as a possible predisposing, accompanying and locally triggering factor.

The 1994 survey confirms the continuing deteriorating trend in forest condition observed on a large geographic scale. This trend cannot be readily explained by site conditions and natural damaging agents alone, even though prolonged drought is thought to contribute to forest decline. Long-range transboundary air pollution is, however, expected to reveal itself through such an effect as deteriorating forest condition. This phenomenon clearly deserves particular attention.

Because of the importance of summer droughts and warm winters for the recent changes in forest condition, special attention should also be paid in the future to the impact of global climate change besides the effects of air pollution.

IV. CONCLUSIONS AND RECOMMENDATIONS

The objective of the joint large-scale monitoring of UN/ECE and EU is to gain knowledge of the spatial and temporal variation of forest condition. This is achieved by means of an annual crown condition assessment in Europe.

Over a period of nine years the forest condition survey has gradually been extended and today it covers a large proportion of the European region. Notably over the last few years a steady increase in defoliation has been observed in several regions in central and eastern Europe and parts of the Mediterranean region. In these regions there is an overall deterioration of the most common species, while in other regions the trends vary more according to species.

One objective of the level I approach has been fulfilled by providing a more comprehensive knowledge of the extent, dynamics and spatial distribution of forest damage in Europe. At the same time a comprehensive database has been created and impetus has been given to forest damage research and environmental policies. These successive inventories by themselves do not allow cause-and-effect relationships to be established. Notably the role of air pollution is difficult to separate from the influence of other environmental factors.

Nevertheless, it is recommended that the large-scale forest condition assessment should continue on an annual basis. In fact, the results of this survey, within a short time and with reasonable effort, give a good overview of the actual forest condition. The continuation of the survey is also justified because the deterioration of forest condition is a source of concern, as no satisfactory explanation has yet been found.

The continuation of the database will make time series analysis and more complex studies linking forest condition with various factors, including air pollution, possible. Lastly, it will make it possible in the long term to assess the effectiveness of air pollution abatement measures.

Also, the yearly evaluation and the reporting of the results are needed in order to keep resource managers and policy makers informed of the forest health status and trends. The way in which future reporting will be carried out is still under discussion. In future reports topics of special interest such as the intensive monitoring on level II will be dealt with.

A synoptic evaluation of the different data sets on level I is needed at the national and European level in order to gain deeper insight into potential relationships between forest damage and air pollution. The complete level I data set can be used for further synoptic analyses together with the data sets of other ICPs running under the Working Group on Effects. Such a study is being planned for the near future and its results will be presented in an overview report.

In addition to the level I network, intensive monitoring was designed and implemented on a smaller number of plots (level II). This level II programme aims at

recognizing factors and processes with special regard to the impact of air pollutants. This type of approach enables correlations to be established between the variation of environmental factors and the reaction of ecosystems. Such correlations may be used to support the determination of critical loads and levels and their exceedances. The data obtained will also lead to a better interpretation of the findings derived from the large-scale systematic network (level I).

Therefore it is recommended that both levels should be maintained with more emphasis on level II.

Annex I
FORESTS AND SURVEYS IN EUROPEAN COUNTRIES (1994)

Participating countries	Total area (1000 ha)	Forest area (1000 ha)	Coniferous forest (1000 ha)	Broadleav. forest (1000 ha)	Area surveyed (1000 ha)	Grid size (km x km)	No. of sample plots	No. of sample trees
Austria	8385	3857	2922	935	3857	8.7 x 8.7	216	6397
Belarus	20760	7028	4757	2271	6001	16 x 16	407	9788
Belgium	3057	602	302	300	602	$8^2 / 16^2$	106	2487
Bulgaria	11100	3314	1172	2142	3314	$16^2 / 8^2$	166	6625
Croatia	5654	2061	321	1740	1175	16 x 16	89	2174
Czech Republic	7886	2630	2051	579	2630	$8^2/16^2$	213	14342
Denmark	4300	466	308	158	411	$7^2/16^2$	54	1296
Estonia	4510	1815	1135	680	1135	16 x 16	91	2184
Finland	30464	20059	18484	1575	15304	$16^2 / 24x32$	381	4261
France	54919	14002	5040	8962	13100	$16^2 /16x1$	534	10672
Germany	35562	10207	6858	3349	10207	4 x 4	8034	219657
Greece a)	13204	2034	954	1080	2034	16 x 16	80	1888
Hungary	9300	1713	266	1447	1600	4 x 4	1064	22304
Ireland	6889	380	334	46	285	16 x 16	22	441
Italy	30126	8675	1735	6940	7699	16 x 16	211	5854
Latvia	6450	2797	1633	1164	2661	8 x 8	370	9154
Liechtenstein	16	8	6	2	no survey in 1994			
Lithuania	6520	1823	1073	750	1823	16 x 16	73	1761
Luxembourg	259	89	32	54	86	$16^2 / 4^2$	51	1169
Rep. of Moldova	3050	1141		1141	1141	2 x 2	571	21453
Netherlands	4147	311	208	103	281	1 x 1	1259	31475
Norway	30686	13700	7000	6700	13700	$9^2/18^2$	911	8412
Poland	31270	8654	6895	1759	8654	16 x 16	1398	27780
Portugal	8800	3370	1338	2032	3370	16 x 16	147	4410
Romania	23750	6244	1929	4315	6244	2x2/2x4	7226	184396
Russian Fed. b)	8530	5798	3972	1826	5798	varying	124	2934
Slovak Republic	4901	1885	816	1069	1185	16 x 16	111	4324
Slovenia	2008	1071	500	571	1071	16 x 16	34	816
Spain	50471	11792	5637	6155	11792	varying	444	10656
Sweden	40800	23500	19729	3771	20009	varying	4675	15080
Switzerland	4129	1186	818	368	1186	8 x 8	164	1958
Turkey	77945	20199	9426	10773	no survey in 1994			
Ukraine	60370	6151	2931	3220	2021	16 x 16	146	3469
United Kingdom	24100	2200	1550	650	2200	random	367	8808
Yugoslavia c)	25600	6100	900	5200	no survey in 1994			
TOTAL	659918	196862	113032	83827	152576	varying	29739	648425

a) Excluding maquis.   b) Only Leningrad Region.
c) Former Yugoslavia excluding Croatia and Slovenia.

Annex II

DEFOLIATION OF ALL SPECIES BY CLASSES AND CLASS AGGREGATES (1994)

Participating countries	Area surveyed (1000 ha)	No. of sample trees	0 none	1 slight	2 moderate	3+4 severe and dead	2+3+4
Austria	3857	6397	59.9	32.3	7.1	0.7	7.8
Belarus	6001	9788	15.6	47.0	35.2	2.2	37.4
Belgium	602	2487	43.0	40.1	15.6	1.3	16.9
Bulgaria	3314	6625	31.8	39.3	25.1	3.8	28.9
Croatia	1175	2174	46.2	25.0	24.6	4.2	28.8
Czech Republic	2630	14342	8.7	31.6	53.8	5.9	59.7
Denmark	466	1296	38.7	24.8	28.9	7.6	36.5
Estonia	1135	2184	46.4	37.9	14.6	1.1	15.7
Finland	15304	4261	63.2	23.8	11.9	1.1	13.0
France	13100	10672	72.4	19.2	7.3	1.1	8.4
Germany	10207	8034	36.3	39.3	22.9	1.5	24.4
Greece a)	2034	1888	38.0	38.8	18.5	4.7	23.2
Hungary	1600	22304	41.9	36.4	15.8	5.9	21.7
Ireland	285	441	only conifers assessed				
Italy	7699	5854	56.6	23.9	16.0	3.5	19.5
Latvia	2661	9154	24.0	46.0	27.0	3.0	30.0
Liechtenstein			no survey in 1994				
Lithuania	1823	1761	14.8	59.8	23.5	1.9	25.4
Luxembourg	86	1169	33.2	32.0	31.0	3.8	34.8
Rep. of Moldova	1141	21453	only broadleaves assessed				
Netherlands	281	31475	60.7	19.9	16.5	2.9	19.4
Norway	13700	8412	37.9	34.6	22.4	5.1	27.5
Poland	8654	27780	5.2	39.9	51.9	3.0	54.9
Portugal	3370	4410	63.8	30.5	5.4	0.3	5.7
Romania	6244	184396	47.7	31.1	18.1	3.1	21.2
Russian Fed. b)	5798		only conifers assessed				
Slovak Republic	1185	4324	14.7	43.5	36.2	5.6	41.8
Slovenia	1071	816	41.0	43.0	13.0	3.0	16.0
Spain	11792	10656	38.4	42.2	13.0	6.4	19.4
Sweden	20009	15080	only conifers assessed				
Switzerland c)	1186	1958	31.8	45.6	20.0	2.6	22.6
Turkey			no survey in 1994				
Ukraine	2021	3469	22.5	45.1	30.0	2.4	32.4
United Kingdom	2200	8808	42.8	43.3	13.0	0.9	13.9
Yugoslavia d)			no survey in 1994				

a) Excluding maquis. b) Only Leningrad Region.
c) Weighted according to diameter breast height (dbh).
d) Former Yugoslavia excluding Croatia and Slovenia.

Annex III

DEFOLIATION OF CONIFERS BY CLASSES AND CLASS AGGREGATES (1994)

Participating countries	Coniferous forest (1000 ha)	No. of sample trees	0 none	1 slight	2 moderate	3+4 severe and dead	2+3+4
Austria a)	2922	5570	61.3	30.8	7.2	0.7	7.9
Belarus	4122	7207	9.6	46.4	41.8	2.2	44.0
Belgium	302	1218	34.7	44.1	19.0	2.2	21.2
Bulgaria	1172	3912	38.9	36.1	22.3	2.7	25.0
Croatia	321	407	43.3	17.4	34.1	5.2	39.3
Czech Republic	2051	12559	8.1	30.7	55.5	5.7	61.2
Denmark	308	827	40.0	21.3	27.6	11.1	38.7
Estonia	1135	2089	44.0	40.0	15.0	1.0	16.0
Finland	18484	3636	63.2	23.7	12.0	1.1	13.1
France	5040	3710	75.2	16.6	7.3	0.9	8.2
Germany	6946	6858	39.4	39.0	20.3	1.3	21.6
Greece b)	954	1016	47.0	39.8	9.9	3.3	13.2
Hungary	248	3590	47.0	31.8	14.3	6.9	21.2
Ireland	334	441	32.9	47.4	19.0	0.7	19.7
Italy	1735	1321	60.4	24.6	11.4	3.6	15.0
Latvia	1606	6756	19.0	47.0	31.0	3.0	34.0
Liechtenstein	6		no survey in 1994				
Lithuania	1073	1207	13.2	60.5	24.5	1.8	26.3
Luxembourg	32	415	56.4	30.8	11.1	1.7	12.9
Rep. of Moldova			only broadleaves assessed				
Netherlands	182	20600	54.4	17.9	24.1	3.6	27.7
Norway	7000	6680	43.5	34.1	18.1	4.3	22.4
Poland	6895	23400	4.9	39.5	52.7	2.9	55.6
Portugal	1338	1652	70.8	23.8	5.3	0.1	5.4
Romania	1929	40787	54.6	29.9	13.6	1.9	15.5
Russian Fed. c)	3972	2934	51.1	41.3	7.0	0.6	7.6
Slovak Republic	816	1799	8.2	41.5	43.5	6.8	50.3
Slovenia	500	371	32.0	49.0	17.0	2.0	19.0
Spain	5637	5394	44.2	36.2	13.3	6.3	19.6
Sweden	19729	15080	57.1	26.7	12.6	3.6	16.2
Switzerland d)	818	1229	29.1	44.9	23.0	3.0	26.0
Turkey	9426		no survey in 1994				
Ukraine	2931	1767	19.9	45.3	32.8	2.0	34.8
United Kingdom	1550	5352	42.1	42.9	13.8	1.2	15.0
Yugoslavia e)	900		no survey in 1994				

a) Only trees 60 years and older assessed. b) Excluding maquis.
c) Only Leningrad Region. d) Weighted according to diameter breast height (dbh).
e) Former Yugoslavia excluding Croatia and Slovenia.

Annex IV

DEFOLIATION OF BROADLEAVES BY CLASSES
AND CLASS AGGREGATES (1994)

Participating countries	Broadleav. forest (1000 ha)	No. of sample trees	0 none	1 slight	2 moderate	3+4 severe and dead	2+3+4
Austria a)	935	827	50.6	41.8	6.5	0.9	7.4
Belarus	1879	2581	32.3	49.1	16.7	1.9	18.6
Belgium	300	1269	51.0	36.2	12.4	0.4	12.8
Bulgaria	2142	2713	21.6	44.0	29.2	5.2	34.4
Croatia	1740	1767	46.8	26.8	22.4	4.0	26.4
Czech Republic	579	1783	13.2	37.9	42.4	6.5	48.9
Denmark	158	469	36.5	31.1	31.1	1.3	32.4
Estonia	680	95	95.0	3.0	1.0	1.0	2.0
Finland	1100	625	63.7	24.3	11.2	0.8	12.0
France	8962	6962	71.0	20.6	7.3	1.1	8.4
Germany	3349	68412	29.8	40.1	28.4	1.7	30.1
Greece b)	1080	872	27.4	37.6	28.5	6.5	35.0
Hungary	1351	18714	40.9	37.3	16.1	5.7	21.8
Ireland	46	only conifers assessed					
Italy	6940	4533	55.5	23.8	17.3	3.4	20.7
Latvia	1055	2398	40.0	45.0	12.0	3.0	15.0
Liechtenstein	2	no survey in 1994					
Lithuania	750	554	18.4	58.3	21.1	2.2	23.3
Luxembourg	54	754	20.5	32.7	43.0	3.8	46.8
Rep. of Moldova	1141	21453	63.5	14.6	18.2	3.7	21.9
Netherlands	99	11875	71.6	23.3	3.4	1.7	5.1
Norway c)	6700	1732	16.4	36.0	39.1	8.5	47.6
Poland	1759	4380	6.9	41.6	48.5	3.0	51.5
Portugal	2032	2758	59.6	34.6	5.4	0.4	5.8
Romania	4315	143609	45.6	31.5	19.4	3.5	22.9
Russian Fed. d)	1826	only conifers assessed					
Slovak Republic	1069	2525	19.3	45.1	31.0	4.6	35.6
Slovenia	571	445	49.0	38.0	10.0	3.0	13.0
Spain	6155	5262	32.5	48.2	12.8	6.5	19.3
Sweden c)	3771	only conifers assessed					
Switzerland e)	368	729	37.0	46.8	14.3	1.9	16.2
Turkey	10773	no survey in 1994					
Ukraine	3220	1702	25.2	44.9	26.9	3.0	29.9
United Kingdom	650	3456	43.8	43.8	11.9	0.5	12.4
Yugoslavia f)	5200	no survey in 1994					

a) Only trees 60 years and older assessed.   b) Excluding maquis.   c) Special study on birch.
d) Only Leningrad Region.   e) Weighted according to diameter breast height (dbh).
f) Former Yugoslavia excluding Croatia and Slovenia.

Annex V

DEFOLIATION OF ALL SPECIES (1986-1994)

Participating countries	All species Defoliation classes 2-4									% change
	1986	1987	1988	1989	1990	1991	1992	1993	1994	1993/1994
Austria				10.8	9.1	7.5	6.9	8.2	7.8	-0.4
Belarus				67.2	54.0		19.2	29.3	37.4	8.1
Belgium				14.6	16.2	17.9	16.9	14.8	16.9	2.1
Bulgaria	8.1	3.6	7.4	24.9	29.1	21.8	23.1	23.2	28.9	5.7
Croatia							15.6	19.2	28.8	9.6
Czech Republic							56.4	53.0	59.7	6.7
Denmark		23.0	18.0	26.0	21.2	29.9	25.9	33.4	36.5	3.1
Estonia	only conifers assessed						28.5	20.3	15.7	-4.6
Finland		12.1	16.1	18.0	17.3	16.0	14.5	15.2	13.0	-2.2
France a)	8.3	9.7	6.9	5.6	7.3	7.1	8.0	8.3	8.4	0.1
Germany b)	18.9	17.3	14.9	15.9	15.9	25.2	26.0	24.2	24.4	0.2
Greece c)			17.0	12.0	17.5	16.9	18.1	21.2	23.2	2.0
Hungary			7.5	12.7	21.7	19.6	21.5	21.0	21.7	0.7
Ireland	only conifers assessed									
Italy				9.1	14.8	16.4	18.2	17.6	19.5	1.9
Latvia					36.0		37.0	35.0	30.0	-5.0
Liechtenstein	19.0	19.0	17.0	11.8			16.0			
Lithuania			3.0	21.5	20.4	23.9	17.5	27.4	25.4	-2.0
Luxembourg	5.1	7.9	10.3	12.3		20.8	20.4	23.8	34.8	11.0
Rep. of Moldova								50.8		
Netherlands	23.3	21.4	18.3	16.1	17.8	17.2	33.4	25.0	19.4	-5.6
Norway	only conifers assessed			18.2	19.7	26.2	24.9	27.5		2.6
Poland			20.4	31.9	38.4	45.0	48.8	50.0	54.9	4.9
Portugal			1.3	9.1	30.7	29.6	22.5	7.3	5.7	-1.6
Romania						9.7	16.7	20.5	21.2	0.7
Russian Fed.	only conifers assessed									
Slovak Republic			38.8	49.2	41.5	28.5	36.0	37.6	41.8	4.2
Slovenia				22.6	18.2	15.9		19.0	16.0	-3.0
Spain			6.8	4.5	4.6	7.3	12.3	13.0	19.4	6.4
Sweden	only conifers assessed									
Switzerland	12.0	15.0	12.0	12.0	17.0	19.0	16.0	18.0	22.6	4.6
Turkey										
Ukraine							6.4	16.3	21.5	32.4	10.9
United Kingdom d)		22.0	25.0	28.0	39.0	56.7	58.3	16.9	13.9	-3.0
Yugoslavia e)						9.8				

a) 16x16 km network after 1988. b) For 1986-1990, only data for former Federal Republic of Germany.
c) Excluding maquis. d) The difference between 1992 and subsequent years is mainly due to a change of assessment method in line with that used in other States.
e) Former Yugoslavia; Croatia and Slovenia excluded from 1991 results.

Annex VI

DEFOLIATION OF CONIFERS (1986-1994)

Participating countries	Conifers Defoliation classes 2-4									% change 1993/1994
	1986	1987	1988	1989	1990	1991	1992	1993	1994	
Austria				10.1	8.3	7.0	6.6	8.2	7.9	-0.3
Belarus				76.0	57.0		33.7	33.8	43.0	9.2
Belgium				20.4	23.6	23.4	23.0	18.3	21.2	2.9
Bulgaria	4.7	3.8	7.6	32.9	37.4	26.5	25.5	26.9	25.0	-1.9
Croatia							26.3	33.9	39.3	5.4
Czech Republic							58.4	52.7	61.2	8.5
Denmark		24.0	21.0	24.0	18.8	31.4	28.6	37.0	38.7	1.7
Estonia			9.0	28.5	20.0	28.0	29.5	21.2	16.0	-5.2
Finland		13.5	17.0	18.7	18.0	17.2	15.2	15.6	13.1	-2.5
France a)	12.5	12.0	9.1	7.3	6.6	6.7	7.1	8.2	8.2	0.0
Germany b)	19.5	15.9	14.0	13.2	15.0	24.8	23.8	21.4	21.6	0.2
Greece			7.7	6.7	10.0	7.2	12.3	13.9	13.2	-0.7
Hungary			9.4	13.3	23.3	17.8	20.1	20.1	21.2	1.1
Ireland		0.0	4.8	13.2	5.4	15.0	15.7	29.6	19.7	-9.9
Italy				9.2	12.8	13.8	17.2	15.1	15.0	-0.1
Latvia					43.0		45.0	41.0	34.0	-7.0
Liechtenstein	22.0	27.0	23.0	12.4			18.0			
Lithuania			3.0	24.0	22.9	27.8	17.5	29.2	26.3	-2.9
Luxembourg	4.2	3.8	11.1	9.5			6.3	9.0	12.8	3.8
Rep. of Moldova								45.2		
Netherlands	28.9	18.7	14.5	17.7	21.4	21.4	34.7	30.6	27.7	-2.9
Norway			20.8	14.8	17.1	19.0	23.4	20.9	22.4	1.5
Poland			24.2	34.5	40.7	46.9	50.3	52.5	55.6	4.8
Portugal			1.7	9.8	25.7	19.8	11.3	7.1	5.4	-1.7
Romania						6.9	10.9	16.6	15.5	-1.1
Russian Fed. c)						4.2	5.2	4.5	7.6	3.1
Slovak Republic			52.7	59.1	55.5	38.5	44.0	49.9	50.3	0.4
Slovenia					34.6	31.3		27.0	19.0	-8.0
Spain			7.7	4.7	4.4	7.2	13.5	14.7	19.6	4.9
Sweden		5.6	12.3	12.9	16.1	12.3	16.9	10.6	16.2	5.6
Switzerland	14.0	16.0	14.0	18.0	20.0	24.0	19.0	20.0	26.0	6.0
Turkey										
Ukraine				1.4	3.0	6.4	13.8	21.7	34.8	13.1
United Kingdom d)		23.0	27.0	34.0	45.0	51.5	52.7	16.8	15.0	-1.8
Yugoslavia e)	23.0	16.1	17.5	39.1	34.6	15.9				

a) 16x16 km network after 1988.   b) For 1986-1990, only data for former Federal Republic of Germany.
c) For 1993-1994, only data for Leningrad Region.   d) The difference between 1992 and subsequent years is mainly due to a change of assessment method in line with that used in other States.   e) Former Yugoslavia; Croatia and Slovenia excluded from 1991 results.

Annex VII

DEFOLIATION OF BROADLEAVES (1986-1994)

Participating countries	Broadleaves Defoliation classes 2-4									% change
	1986	1987	1988	1989	1990	1991	1992	1993	1994	1993/1994
Austria				15.7	14.9	11.1	9.3	7.7	7.4	-0.3
Belarus				33.4	45.0		14.8	16.6	18.6	2.0
Belgium				8.7	10.0	13.5	11.8	11.7	12.8	1.1
Bulgaria	4.0	3.1	8.8	16.2	17.3	15.3	18.0	16.6	34.4	17.8
Croatia							13.6	15.6	26.4	10.8
Czech Republic							31.9	55.1	48.9	-6.2
Denmark		20.0	14.0	30.0	25.4	27.3	21.2	27.0	32.4	5.4
Estonia				only conifers assessed				1.1	2.0	0.9
Finland		4.7	7.9	12.6	11.6	7.7	10.1	12.8	12.0	-0.8
France ·a)	4.8	6.5	5.3	4.8	7.7	7.4	8.5	8.4	8.4	0.0
Germany b)	16.8	19.2	16.5	20.4	23.8	26.5	32.0	29.9	30.1	0.2
Greece			28.5	18.4	26.5	28.5	25.0	29.8	35.0	5.2
Hungary			7.0	12.5	21.5	19.9	21.8	21.2	21.8	0.6
Ireland				only conifers assessed						
Italy		3.6	2.9	9.5	15.4	17.1	18.5	18.3	20.7	2.4
Latvia					27.0		19.0	17.8	15.0	-2.8
Liechtenstein	10.0	7.0	5.0	9.0			8.0			
Lithuania			1.0	16.0	15.8	14.9	17.6	23.8	23.3	-0.5
Luxembourg	5.6	10.1	12.3	13.9		33.9	30.5	31.0	46.8	15.8
Rep. of Moldova								50.9	21.9	-29.0
Netherlands	13.2	26.5	25.4	13.1	11.5	9.4	31.1	13.1	5.1	-8.0
Norway					18.2	25.1	38.9	42.1	47.6	5.5
Poland			7.1	17.7	25.6	34.8	40.4	49.9	51.5	5.9
Portugal			0.8	8.6	34.1	36.6	29.1	7.5	5.8	-1.7
Romania						10.4	18.4	21.4	22.9	1.5
Russian Fed.				only conifers assessed						
Slovak Republic			28.5	41.8	31.3	21.1	30.0	29.1	35.6	6.5
Slovenia					4.4	5.8		11.0	13.0	2.0
Spain			7.4	4.2	4.8	7.4	11.2	11.4	19.3	7.9
Sweden				only conifers assessed						
Switzerland	8.0	13.0	6.0	5.0	13.0	15.0	11.0	13.0	16.2	3.2
Turkey										
Ukraine				1.4	2.7	6.5	20.2	21.6	29.9	8.3
United Kingdom c)		20.0	20.0	21.0	28.8	65.6	67.8	17.1	12.4	-4.7
Yugoslavia d)		7.3	9.0	8.2	4.4	8.2				

a) 16x16 km network after 1988.   b) For 1986-1990, only data for former Federal Republic of Germany.
c) The difference between 1992 and subsequent years is mainly due to a change of assessment method in line with that used in other States.   d) Former Yugoslavia: Croatia and Slovenia excluded from 1991 results.

Part THREE

CALCULATION OF CRITICAL LOADS OF NITROGEN AS A NUTRIENT

Summary Report on the Development of a Library of Default Values

INTRODUCTION

The workshop on calculation and mapping of critical loads of nitrogen held from 24 to 26 October 1994 at Grange-over-Sands (United Kingdom) proposed the following form of the simple mass balance equation :

$$CL_{(N)nut} = N_{l(crit)} + N_{i(crit)} + N_{u(crit)} + N_{de(crit)} \\ -N_{fix(crit)} + N_{fire(crit)} + N_{erode(crit)} + N_{vol(crit)}$$

where:

$CL_{(N)nut}$ = critical load for nutrient nitrogen;

$N_{l(crit)}$ = total annual N leaching (NO_3 + NH_4 - dissolved organic N) from the rooting zone under natural conditions in the absence of pollutant N inputs plus any enhanced leaching following forest harvesting, natural fires or fires used as part of traditional management regimes;

$N_{i(crit)}$ = an acceptable annual level of N immobilization in soil organic matter (including forest floor), at N inputs equal to the critical load, at which adverse ecosystem change will not take place;

$N_{u(crit)}$ = net annual removal of N in vegetation and harvested animals at N inputs equal to the critical load;

$N_{de(crit)}$ = annual flux of N to the atmosphere as a result of denitrification at N inputs equal to the critical load;

$N_{fix(crit)}$ = annual N fixation;

$N_{fire(crit)}$ = N losses in smoke from natural wildfires or from fires used as part of traditional management regimes;

$N_{erode(crit)}$ = annual N losses through erosion under natural conditions plus enhanced erosion losses following forest harvesting, natural fires or fires used as part of traditional management regimes;

$N_{vol(crit)}$ = annual N losses to the atmosphere from NH_4 volatilization.

National databases of measured values of the various input variables will rarely be available. As a result, recommended default values are likely to be used as input in the production of provisional national critical load maps. The workshop held at Lokeberg (Sweden) in 1993[1] suggested a number of default values but also recommended that published data be brought together to provide the basis for the development of a matrix of values for the variables and a range of major ecosystem types. The workshop at Grange-over-Sands reiterated the need for such a review of available data. As a result of the recommendations of the workshop, the establishment of a database of default values was initiated by the joint effort of the United Kingdom Department of the Environment and the ECE secretariat.

I. OBJECTIVES AND METHOD OF WORK

The objectives of the work can be summarized as follows:

(a) To gather and critically assess available data quantifying the input variables to the simple mass balance equation for the calculation of critical loads of nitrogen as a nutrient;

(b) To develop a library of default values for each of the input variables to the simple mass balance equation and for a range of ecosystems.

Literature searches have been carried out, using computerized bibliographic databases in the United Kingdom and the United States of America, for combinations of relevant key words for each of the input variables to the simple mass balance equation. The resulting listings of titles were then searched manually to select subsets of relevant publications. Abstracts were then obtained from the bibliographies for the selected, relevant papers. Where necessary, copies of the full text of the relevant papers were obtained.

In parallel with the above procedure, a manual literature search was carried out (i) to check the recovery by

---

[1] Grennfelt P. and E. Thornelof (eds.), *Critical loads for nitrogen* (Copenhagen, Nordic Council of Ministers, 1992).

the key word combinations used in the searches of the computerized bibliographies and (ii) to access selected papers cited in publications identified from the computer searches.

II. DATA PRESENTATION

A. Data listing for each variable

The values for each input variable are being entered into a computer database, which also includes, where available, the following information: site name; site location (country and county, State, Land, district, etc.); latitude and longitude; altitude; area of site (where site is a stream catchment); habitat/ecosystem type (using descriptors employed by the authors of the publication); metrology used to obtain the measurements; source of data (reference); general comments.

B. Summary data

A summary data file is being produced which contains: (*a*) the range of values for each variable—ecosystem combination; and (*b*) mean/recommended value for each variable—ecosystem combination.

C. Mapping manual

A condensed table and/or guidance for the calculation of default values will be prepared for inclusion in the mapping manual.

III. PROGRESS MADE AND RESULTS ACHIEVED

A large variation has been found in the number of published values recovered to date for the different variables. For example, there are more than 200 reliable values for N_l but, on the basis of current searches, only one or two reliable values for N_{vol} from soils; there are however a large number of values for N_{vol} from animal wastes. Some 40 reliable values for N_{de} have been identified to date. There are very few values for current rates of N_i and mean values are being calculated from data for accumulation of N in soil chronosequences and in disturbed systems, e.g. arable land allowed to revert to forest. (These values only provide extremes which can be used to scale and test other data.) Up to now, surprisingly few values have been found for N_{erode}; there are a large number of values for mass of material lost in erosion, but few publications give the chemical composition of the eroded material. The available data show that erosion losses following perturbations such as fire and clearcutting can be considerable for a number of years. Losses in smoke from fires can also be large; the best method of calculation may be from the proportion of the N content in the standing biomass lost. Net losses in grazing animals from extensively grazed semi-natural ecosystems can also be significant.

There is a clumping in the number of available values for variables in terms of thecountries of origin and ecosystem type. Thus most of the available values for N_{de} originate from the United States, while the bulk of the values for N_l originate from north-west Europe. The available data are also heavily biased towards forest ecosystems, and particularly coniferous forests.

In an attempt to locate further data for those variables for which few values have been found to date, larger bibliographic databases in the United States are now being accessed and relevant contacts have been established with the scientists from eastern Europe.

The results of the work already done on assessing available data quantifying the input variables to the simple mass balance equation for the calculation of critical loads of nitrogen as nutrient, together with some preliminary conclusions, are presented in the following paragraphs.

A. Critical total annual leaching of nitrogen

Available data on total annual leaching of nitrogen from Europe have been brought together by Hauhs *et al.*[2], Hornung *et al.*[3] and Dise and Wright[4]. The present exercise has drawn on these reviews, but has also brought together some data from North America.

The rates of loss are small for undisturbed natural systems but can increase for a number of years following disturbance, e.g. fire, windblow, and felling. The number of years of higher rates of loss depends on the rate of recolonization and this, in turn, depends on site fertility. The total losses over the period of enhanced leaching following disturbance should be divided by the period between disturbance events and the resultant value added to the background level of leaching.

The following ranges of values have been found in the literature:.

    (i) Boreal and temperate heaths and bogs: 0-0.5 kg ha^{-1} yr^{-1} (inorganic N); losses of organic N can be larger and data are being accumulated for these outputs. (There is, however, an urgent need for more data on organic N outputs from a range of ecosystems);

    (ii) Managed coniferous forest: 0.5-1.0 kg ha^{-1} yr^{-1};

    (iii) Coniferous plantations: 1-3 kg ha^{-1} yr^{-1}. (Can be significantly larger if open drains are dug prior to planting);

---

[2] Hauhs M. and others, "Summary of European data", in *The Role of Nitrogen in the Acidification of Soils and Surface Waters*, edited by J. L. Malanchuk and J. Nilsson (Copenhagen, Nordic Council of Ministers, 1989).

[3] Hornung M., P. Roda and S. J. Langan, "A review of small catchment studies in Western Europe producing hydrochemical budgets", *Air Pollution Research Report*, 28 (Brussels, Commission of the European Communities, 1990).

[4] Dise N. B. and R. F. Wright, "Nitrogen leaching from European forests in relation to nitrogen deposition", *Forest Ecology and Management*, 71 (1995) pp. 153-161.

(iv) Temperate deciduous forests: 2-4 kg ha^{-1} yr^{-1};

(v) Temperate grasslands: 1-3 kg ha^{-1} yr^{-1};

(vi) Mediterranean forests: 1-2 kg ha^{-1} yr^{-1};

(vii) Temperate forests following felling: 5-25 kg ha^{-1} yr^{-1} for a period of *ca.* two years;

(viii) Heathland following fire: 2-10 kg ha^{-1} yr^{-1};

(ix) Temperate forest following fire: 5-20 kg ha^{-1} yr^{-1}.

B.   Critical annual level of nitrogen immobilization

The annual level of nitrogen immobilization is the most difficult of the input variables to the N mass balance equation to parametrize. There are very few reported values for rates of immobilization but an increasing number of studies are applying ^{15}N techniques to determine rates over the short term.

Rates of N immobilization can be calculated from chronosequence studies. The total amount of N accumulated in a soil is divided by the period of soil formation. Studies of this type on natural ecosystems, usually considering soils developed on glacial moraines or on sand-dunes, indicate rates of 0.5-1.0 kg ha^{-1} yr^{-1}.

Studies of aggrading systems where forest has developed on former agricultural land in temperate areas show values of 3-10 kg ha^{-1} yr^{-1}.

Alternative approaches to the calculation of immobilization rates are being explored as a matter of urgency.

C.   Critical annual removal of nitrogen

Methods of calculation of net uptake by vegetation are outlined in the mapping manual. One method for forests divides the N in biomass by the age of the forest; in the absence of site-specific or national databases, the large amount of published data dealing with nitrogen losses in smoke from fires, N_{fire}, can be used. Net uptake from non-forest, natural or semi-natural ecosystems is insignificant unless the systems are used as extensive grazings for sheep, goats or cattle (or reindeer?). Net removal of sheep from extensive sheep grazings in the United Kingdom is between 0.5 and 2 kg ha^{-1} yr^{-1}, depending on site fertility and grazing densities.

D.   Critical annual flux of nitrogen to the atmosphere

The annual flux of nitrogen to the atmosphere can be calculated using approaches developed by Sverdrup and Ineson and by de Vries. In the absence of the necessary input data required by these methodologies, a default value can be used. Dutch and Ineson[5] reviewed the avail-

able data on rates of denitrification and the present exercise used their study as a starting-point. Some 40 separate values have been collected from the literature, mainly for North American sites but also covering a range of European studies. The studies include both direct field measurements of denitrification, usually for short intervals, which are then extrapolated to longer periods, and laboratory measurements. They also include measurements of actual rates and potential rates. The latter values are generally higher than actual measured field rates.

Losses of N by denitrification are generally small in undisturbed natural systems with low background levels of N inputs. However, rates of denitrification can increase following disturbance. The increased rates will last for only one to three years. The total increased losses following the disturbance should be divided by the period of time between disturbance events and added to the background rates.

The following values have been considered:

(i) Boreal and temperate ecosystems: 0.1 to 3.0 kg ha^{-1} yr^{-1} with the majority of reported values, especially for forest systems, below 1 kg ha^{-1}yr^{-1}; the upper values apply to wet soils; rates for well-drained soils are generally less than 0.5 kg ha^{-1} yr^{-1};

(ii) Following felling: 2.0 to 30.0 kg ha^{-1} yr^{-1} for a two-year period following felling; the upper values apply to nutrient-rich sites with large nitrate concentrations in soil solution, the lower to acid soils with low nitrate concentrations. The value of 30 kg ha^{-1} yr^{-1} is a potential rate.

E.   Critical nitrogen losses in smoke from fires

Nitrogen losses in smoke and gases are only significant in a few natural ecosystems (e.g. in maquis and boreal forests), and in semi-natural systems where fire is a normal part of management used to maintain the ecosystem (e.g. Calluna heathland in the United Kingdom). (Fire is also being used increasingly in boreal and temperate forest management in North America to maintain diversity.)

There are published data on outputs at fire but examination of the data suggests that the losses should be calculated from the N content of the above ground biomass and litter layers. Between 60 and 80 per cent of the N will be lost, the percentage varying with the temperature of the fire; *ca.* 60 per cent losses from "cool" fires and *ca.* 80 per cent losses from "hot" fires. There may also be some loss of soil N from the surface organic horizons in the hot fires. The losses in fires should be divided by the time interval between the occurrence of the fires. Return time for fires in Mediterranean maquis vegetation is *ca.* 20 years (R. Guardans, pers. comm.) and for boreal forest 100-200 years.

There are many published data on the N content in biomass of different ecosystems and forest types and a sample of these will be included in the database. The

---

[5] Dutch J. and P. Ineson, "Denitrification of an upland forest site", *Forestry*, 63 (1990) pp. 363-377.

values below give an indication of the median content and the range of contents in major forest types:

(i) Boreal forest: 447 kg ha^{-1} (174-1,915 kg ha^{-1});

(ii) Temperate coniferous: 664 kg ha^{-1} (375-1,327 kg ha^{-1});

(iii) Temperate broad-leaved: 1,085 kg ha^{-1} (406-1,608 kg ha^{-1});

(iv) Fire losses from Calluna heathland of 20-45 kg ha^{-1} have been reported.

F.   Critical annual nitrogen losses through erosion

There are relatively few published rates of N loss by erosion but many more data on total mass loss and these data can be used to calculate approximate rates of N loss.

Losses by erosion are very small from undisturbed natural ecosystems but these losses can increase dramatically following disturbance, e.g. windblow, fire, clearcutting. The increased rates of loss will last for only a small number of years after disturbance, the length of time depending on the rates of recolonization by vegetation: this is rapid, up to one year, on nutrient-rich sites and slow, two to five years (or more in tundra systems), on nutrient-poor sites. The total losses resulting from the period of increased erosion following disturbance should be divided by the time interval between disturbance events and the resultant value added to the background rate of loss.

Ranges of losses following disturbance are listed below for a few ecosystems:

(i) Undisturbed natural systems: < 0.5 kg ha^{-1} yr^{-1};

(ii) Temperate plantations: 1-3 kg ha^{-1} yr^{-1};

(iii) Heathland following fire: 80-150 kg ha^{-1} yr^{-1} for a period of one to three years; the variation is related to differences in slope and rainfall;

(iv) Boreal and temperate forest following fire: 50-200 kg ha^{-1} yr^{-1}; the rate for a given site is determined by fire intensity, slope and climate (mainly rainfall amount and intensity);

(v) Temperate forest following clearcutting: 5-10 kg ha^{-1} yr^{-1} for a period of two to five years.

Part FOUR

EUROPEAN SULPHUR AND NITROGEN EMISSIONS, DEPOSITIONS FOR 1980 AND 1993 AND EXPORT/IMPORT BUDGETS

In compliance with the work-plan for the implementation of the Convention (ECE/EB.AIR/42, annex I, section 2.2), model calculations for the long-range transmission of sulphur and nitrogen compounds are carried out on a continuous basis.

In the presentation of modelling results, countries are designated by letter codes (table 1). Other areas for calculation are explained in the same table. Tables of deposition of oxidized sulphur and oxidized nitrogen for 1980 and 1993 and of deposition of reduced nitrogen for 1993 and export/import budgets of oxidized sulphur and nitrogen (tables 2 to 8), as well as the total deposition of oxidized sulphur in 1980 and 1990 and change in exceedance of the critical sulphur deposition expressed in figures 1 to 3 have been taken from the report 1/95 EMEP/MSC-W: European Transboundary Acidifying Air Pollution: Ten years of calculated fields and budgets to the end of the first sulphur Protocol.

In deposition (matrix) tables, the columns show calculated depositions from the country or area shown at the top of the column to each country or area in Europe. Horizontal lines refer to depositions in the country shown on the left attributed to emissions from each country in Europe. Each table consists of two parts. Tables 4 and 8 give the sulphur and nitrogen balance in 1980 and in 1993 for each country between emissions lost beyond the national borders (export) and depositions within the country as a consequence of inward transboundary transport (import).

Tables 9 and 10 show the trends in the EMEP sulphur measurements for selected results from the time series analysis. They are the first results of the trend analysis of the sulphur measurements from 1980 to 1993. For sulphur dioxide, significant downward concentration trends were found for measurement sites in Germany, Scandinavia and the United Kingdom. For particulate sulphate the concentration reductions were less pronounced; however, the trends were not increasing anywhere.

NOTE: Names of countries and areas reflect the designations used by MSC-W and do not imply the expression of any opinion whatsoever on the part of the Secretariat of the United Nations concerning the legal status of any country, territory, or area or of its authorities, concerning the delimitation of its frontiers or boundaries.

TABLE 1

Letter codes of countries and other areas

Emitting regions displayed:

Albania	AL	Romania	RO
Austria	AT	Russian Federation,	
Belarus	BY	European part	RU
Belgium	BE	Slovakia	SK
Bosnia and Herzegovina	BA	Slovenia	SI
Bulgaria	BG	Spain	ES
Croatia	HR	Sweden	SE
Czech Republic	CS*	Switzerland	CH
Czechoslovakia	CS	Republic of Moldova	MD
Denmark	DK	Turkey	TR
Estonia	EE	Ukraine	UA
Finland	FI	United Kingdom	GB
France	FR	USSR, European part	SU
Germany, Federal Republic	DE	Yugoslavia, excluding BA,	
Germany, Democratic Republic	DD	HR, FYM, SI	YU*
Greece	GR	Former Yugoslavia	YU
Hungary	HU	Remaining Land Areas	REM
Iceland	IS	Baltic Sea+	BAS
Ireland	IE	Black Sea+	BLS
Italy	IT	Mediterranean Sea+	MED
Latvia	LV	North Sea+	NOS
Lithuania	LI	Remaining N.E.	
Luxembourg	LU	Atlantic Ocean+	ATL
Netherlands	NL	Natural marine sources	NAT
Norway	NO	Kaliningrad region of RU	
Poland	PO	Kola/Karelia region of RU	
Portugal	PT	St. Petersburg/Novgorod-	
The Former Yugoslav		Pskov region of RU	
Republic of Macedonia	FYM	Total inattributable sources	IND
		Total attributable sources	SUM

+ Refers to depositions arising from international trade emissions only.

TABLE 2

Deposition of oxidized sulphur in 1980 (Hundreds of tonnes of S)

Emitters

	AL	AT	BE	BG	CS	DK	FI	FR	DD	DE	GR	HU	IS	IE	IT	LU	NL	NO	PL	
AL	62	2	1	31	11	0	0	9	8	5	23	14	0	0	91	0	0	0	7	AL
AT	1	312	35	8	342	5	1	169	278	327	1	110	0	1	167	2	14	0	135	AT
BE	0	2	539	0	21	3	0	329	49	280	0	3	0	3	4	3	83	0	12	BE
BG	12	9	3	1300	73	2	1	13	51	23	23	112	0	0	54	0	2	0	66	BG
CS	1	104	53	16	2420	12	3	152	1409	443	2	410	0	2	70	3	24	1	775	CS
DK	0	2	23	0	35	195	2	36	104	103	0	6	0	3	2	0	19	4	51	DK
FI	0	3	13	3	60	31	687	23	121	69	0	18	0	2	3	0	9	14	148	FI
FR	0	25	422	3	155	11	1	4314	295	723	1	37	0	20	295	19	116	1	90	FR
DD	0	15	116	3	730	48	2	199	3462	945	0	29	0	4	15	4	66	2	270	DD
DE	0	70	545	5	477	56	2	1090	1165	3780	0	46	0	12	97	24	276	3	207	DE
GR	31	4	2	439	30	1	0	14	23	12	157	38	0	0	113	0	1	0	28	GR
HU	2	82	12	33	453	3	1	50	177	92	3	1483	0	0	114	1	6	0	218	HU
IS	0	0	1	0	1	1	0	3	3	4	0	0	3	1	0	0	1	0	1	IS
IE	0	0	7	0	4	1	0	24	11	15	0	0	0	151	0	0	4	0	4	IE
IT	5	69	25	38	154	4	1	421	146	160	11	118	0	1	2835	1	10	0	104	IT
LU	0	0	9	0	2	0	0	44	5	22	0	0	0	0	1	7	2	0	1	LU
NL	0	2	193	0	28	5	0	167	76	378	0	3	0	4	3	1	318	0	19	NL
NO	0	3	31	1	56	59	25	54	147	123	0	12	0	9	2	1	24	132	84	NO
PL	1	60	113	26	1490	106	15	232	2834	787	2	322	0	5	68	4	65	6	5322	PL
PT	0	0	1	0	1	0	0	11	1	3	0	0	0	1	3	0	1	0	1	PT
RO	8	31	12	327	357	6	4	42	209	88	13	573	0	0	115	1	7	0	333	RO
ES	0	2	22	1	16	2	0	261	32	50	0	7	0	4	59	1	9	0	12	ES
SE	0	7	52	2	141	229	133	85	343	246	0	31	0	7	7	1	38	75	275	SE
CH	0	11	16	1	22	1	0	209	38	95	0	6	0	1	129	1	5	0	11	CH
TR	6	7	4	353	60	2	2	17	50	25	61	65	0	0	66	0	2	0	75	TR
SU	9	99	140	490	1577	206	671	301	1930	826	26	993	0	10	200	4	86	37	3906	SU
GB	0	3	92	1	49	14	2	235	121	169	0	6	0	85	5	1	63	2	43	GB
YU	42	111	19	199	358	5	2	121	239	142	35	588	0	1	616	1	9	0	229	YU
REM	2	6	11	41	42	3	6	120	49	38	13	27	0	1	249	0	5	0	68	REM
BAS	0	16	95	8	319	419	379	159	846	542	0	70	0	8	16	2	67	30	831	BAS
NOS	0	16	468	3	281	230	16	958	750	964	0	36	0	81	22	5	516	71	299	NOS
ATL	0	11	218	3	174	83	129	1050	417	520	0	28	11	392	31	5	122	99	217	ATL
MED	75	99	74	775	421	12	3	1363	382	309	409	396	0	5	4671	4	31	1	329	MED
BLS	5	12	7	735	131	5	2	20	114	48	26	129	0	0	50	0	4	0	214	BLS
SUM	262	1194	3374	4842	10489	1756	2093	12292	15886	12351	804	5717	14	812	10175	95	2003	479	14384	SUM
	AL	AT	BE	BG	CS	DK	FI	FR	DD	DE	GR	HU	IS	IE	IT	LU	NL	NO	PL	

Emitters

	PT	RO	ES	SE	CH	TR	SU	GB	YU	REM	BAS	NOS	ATL	MED	BLS	NAT	IND	SUM	
AL	0	10	3	0	0	2	6	2	38	3	0	0	0	0	0	1	66	397	AL
AT	0	23	17	2	27	0	18	53	99	1	1	3	1	0	0	2	187	2342	AT
BE	0	1	16	1	1	0	2	161	1	0	0	14	2	0	0	3	43	1574	BE
BG	0	369	3	1	1	18	140	7	128	2	0	0	0	0	0	2	205	2617	BG
CS	0	62	18	5	8	1	57	84	64	1	1	5	1	0	0	2	227	6436	CS
DK	0	2	4	16	0	0	12	161	1	0	5	10	1	0	0	5	49	850	DK
FI	0	10	2	114	0	0	747	68	4	0	14	4	1	0	0	8	310	2487	FI
FR	16	9	710	4	41	0	15	594	32	9	1	52	43	0	0	35	729	8817	FR
DD	1	9	19	9	3	0	28	205	8	0	5	14	2	0	0	5	156	6371	DD
DE	3	14	93	9	46	0	34	638	20	1	5	47	8	0	0	13	410	9196	DE
GR	0	75	6	1	0	33	53	5	56	6	0	0	0	0	0	3	211	1341	GR
HU	0	148	8	2	4	2	49	20	154	1	0	1	0	0	0	0	153	3272	HU
IS	0	0	1	1	0	0	2	19	0	0	0	0	1	0	0	15	44	103	IS
IE	0	0	9	0	0	0	1	172	0	0	0	2	9	0	0	16	64	499	IE
IT	3	30	101	2	54	4	20	52	194	26	0	3	2	0	0	10	561	5162	IT
LU	0	0	3	0	0	0	0	5	0	0	0	0	0	0	0	0	6	109	LU
NL	0	1	11	1	1	0	4	296	1	0	0	22	2	0	0	4	47	1587	NL
NO	0	6	8	83	0	0	165	334	2	0	5	15	6	0	0	32	358	1776	NO
PL	1	101	29	47	7	2	311	270	69	1	17	17	3	0	0	8	498	12836	PL
PT	165	0	185	0	0	0	0	6	0	2	0	0	21	0	0	6	65	473	PT
RO	0	2025	8	4	3	17	466	25	246	2	1	2	0	0	0	2	407	5332	RO
ES	105	2	3163	1	2	0	2	79	7	19	0	4	39	2	0	19	384	4306	ES
SE	0	16	8	579	3	1	277	313	7	0	25	18	4	0	0	20	420	3359	SE
CH	1	2	28	0	114	0	2	29	7	0	0	1	1	0	0	1	97	831	CH
TR	0	149	6	2	1	956	324	9	40	14	0	1	0	0	0	6	809	3112	TR
SU	1	1083	45	288	11	162	27060	482	256	147	63	27	6	0	0	38	6465	47640	SU
GB	2	3	44	5	1	0	15	5196	2	0	1	43	27	0	0	41	228	6497	GB
YU	1	199	28	3	8	9	79	36	1486	9	1	2	1	0	0	5	507	5088	YU
REM	7	31	170	3	3	30	1026	30	23	579	1	2	3	1	0	13	1601	4203	REM
BAS	1	32	17	459	3	1	586	422	17	1	108	27	4	0	0	24	477	5984	BAS
NOS	4	13	113	82	4	0	87	5603	9	1	13	282	44	0	0	142	695	11807	NOS
ATL	155	13	2241	111	6	0	1596	2923	8	4	9	71	624	0	0	1340	3975	16588	ATL
MED	26	281	1107	7	36	355	257	175	634	317	2	9	14	2	0	109	2288	14976	MED
BLS	0	394	3	5	1	239	1367	15	69	21	1	1	0	0	0	14	687	4323	BLS
SUM	492	5111	8225	1846	386	1830	34805	18487	3679	1165	277	699	871	6	0	1942	23427	202291	SUM
	PT	RO	ES	SE	CH	TR	SU	GB	YU	REM	BAS	NOS	ATL	MED	BLS	NAT	IND	SUM	

Receivers

TABLE 3

Deposition of oxidized nitrogen in 1980 (Hundreds of tonnes of N)

emitters

	AL	AT	BE	BG	CS	DK	FI	FR	DD	DE	GR	HU	IS	IE	IT	LU	NL	NO	PL	
AL	4	1	0	4	3	0	0	4	1	5	8	2	0	0	22	0	0	0	2	AL
AT	0	44	18	2	73	2	0	77	22	224	1	11	0	0	69	2	16	1	33	AT
BE	0	1	44	0	5	1	0	58	4	68	0	0	0	1	1	1	28	0	3	BE
BG	2	4	2	85	21	1	0	6	5	20	14	14	0	0	16	0	2	0	20	BG
CS	0	32	28	3	267	7	1	70	73	301	1	34	0	1	31	2	31	1	134	CS
DK	0	1	9	0	9	21	0	15	9	53	0	1	0	1	1	0	18	2	12	DK
FI	0	2	9	0	21	21	99	13	16	67	0	3	0	1	2	0	15	17	51	FI
FR	0	11	117	1	43	6	1	930	26	390	0	5	0	6	102	8	82	1	25	FR
DD	0	7	47	1	99	15	1	78	101	394	0	4	0	1	8	3	62	2	53	DD
DE	0	21	155	1	106	16	1	347	75	1060	0	6	0	4	44	11	172	3	50	DE
GR	4	2	1	38	9	0	0	7	2	11	47	5	0	0	25	0	1	0	8	GR
HU	0	25	7	7	87	2	0	24	15	79	2	68	0	0	44	1	8	0	58	HU
IS	0	0	1	0	0	1	0	2	0	4	0	0	2	0	0	0	2	1	1	IS
IE	0	0	4	0	2	1	0	12	1	13	0	0	0	11	0	0	5	0	1	IE
IT	1	25	13	6	46	2	0	184	13	129	7	15	0	0	523	1	12	0	31	IT
LU	0	0	3	0	1	0	0	9	0	8	0	0	0	0	0	1	1	0	0	LU
NL	0	1	28	0	6	2	0	34	5	78	0	0	0	1	1	0	55	0	4	NL
NO	0	2	16	0	18	28	10	27	17	102	0	2	0	4	1	1	31	53	29	NO
PL	0	25	56	5	307	43	6	109	178	515	1	35	0	2	30	3	79	6	593	PL
PT	0	0	1	0	0	0	0	5	0	2	0	0	0	0	1	0	1	0	0	PT
RO	2	14	7	50	94	4	1	20	20	77	8	60	0	0	40	0	10	1	94	RO
ES	0	2	11	0	6	1	0	107	4	41	0	1	0	1	18	1	11	0	4	ES
SE	0	3	27	0	43	70	42	40	37	182	0	4	0	3	3	1	48	52	82	SE
CH	0	3	7	0	6	1	0	83	3	58	0	1	0	0	39	1	6	0	3	CH
TR	1	4	3	42	21	1	1	9	6	25	33	9	0	0	15	0	3	0	26	TR
SU	2	51	83	83	481	133	225	168	219	748	16	125	0	4	83	4	132	49	1110	SU
GB	0	1	27	0	14	7	1	72	11	96	0	1	0	17	2	1	41	2	13	GB
YU	6	39	11	32	102	3	1	58	23	126	19	62	0	0	207	1	12	0	70	YU
REM	1	4	7	7	16	2	3	72	6	40	8	4	0	0	61	0	7	1	25	REM
BAS	0	7	43	2	80	91	53	69	73	319	0	8	0	3	8	2	74	24	171	BAS
NOS	0	6	111	0	65	56	5	238	57	446	0	4	1	21	9	3	199	23	69	NOS
ATL	0	7	104	1	60	55	51	378	50	423	0	4	10	61	12	4	142	67	77	ATL
MED	10	40	36	84	111	6	1	475	32	243	132	44	0	1	745	3	34	1	89	MED
BLS	1	6	4	54	39	4	2	11	12	43	13	15	0	0	14	0	5	1	62	BLS
SUM	33	386	1038	509	2260	601	506	3811	1117	6388	310	546	13	146	2175	52	1344	308	3002	SUM
	AL	AT	BE	BG	CS	DK	FI	FR	DD	DE	GR	HU	IS	IE	IT	LU	NL	NO	PL	

emitters

	PT	RO	ES	SE	CH	TR	SU	GB	YU	REM	BAS	NOS	ATL	MED	BLS	NAT	IND	SUM	
AL	0	2	1	0	0	0	2	1	4	0	0	0	0	0	0	0	21	90	AL
AT	0	4	3	2	24	0	5	29	10	0	1	4	1	0	0	0	73	750	AT
BE	0	0	3	1	1	0	1	61	0	0	0	7	2	0	0	0	19	311	BE
BG	0	42	1	1	1	3	34	3	9	1	0	0	0	0	0	0	60	368	BG
CS	0	10	3	5	9	0	15	48	9	0	2	6	2	0	0	0	94	1219	CS
DK	0	0	1	6	1	0	3	65	0	0	2	6	1	0	0	0	22	261	DK
FI	0	2	0	67	1	0	85	41	1	0	11	5	1	0	0	0	114	664	FI
FR	12	2	142	4	32	0	4	272	6	1	1	34	42	0	0	0	305	2611	FR
DD	0	2	4	7	4	0	7	93	1	0	3	12	2	0	0	0	71	1079	DD
DE	2	2	18	7	37	0	9	267	3	0	3	32	9	0	0	0	175	2633	DE
GR	0	12	2	0	1	5	15	3	5	1	0	0	0	0	0	0	68	274	GR
HU	0	19	2	1	4	0	13	12	19	0	0	2	0	0	0	0	56	554	HU
IS	0	0	0	1	0	0	0	12	0	0	0	1	1	0	0	0	29	59	IS
IE	0	0	0	2	0	0	0	54	0	0	0	3	7	0	0	0	32	148	IE
IT	2	6	29	2	40	1	6	27	26	4	0	3	3	0	0	0	206	1362	IT
LU	0	0	1	0	0	0	0	3	0	0	0	0	0	0	0	0	2	31	LU
NL	0	0	2	1	0	0	0	96	0	0	0	10	2	0	0	0	20	349	NL
NO	0	1	2	43	1	0	15	153	0	0	5	14	5	0	0	0	149	727	NO
PL	1	15	6	32	9	0	70	139	10	0	11	18	4	0	0	0	199	2508	PL
PT	41	0	37	0	0	0	0	4	0	0	0	0	15	0	0	0	42	151	PT
RO	0	137	2	3	3	3	97	15	21	1	1	2	0	0	0	0	124	910	RO
ES	60	1	441	1	3	0	1	44	1	3	0	4	34	2	0	0	186	990	ES
SE	0	3	2	154	2	0	55	149	1	0	16	18	4	0	0	0	165	1204	SE
CH	0	0	6	0	34	0	0	14	1	0	0	2	1	0	0	0	34	304	CH
TR	0	26	1	2	1	77	101	5	5	8	0	1	0	0	0	0	288	713	TR
SU	1	156	11	234	17	26	3776	294	36	83	59	36	8	0	0	0	2168	10619	SU
GB	1	1	7	5	1	0	4	626	0	0	1	23	20	0	0	0	106	1104	GB
YU	0	31	7	3	10	1	21	19	79	1	1	2	1	0	0	0	170	1117	YU
REM	5	6	53	3	5	5	237	20	4	107	1	2	4	1	0	0	859	1576	REM
BAS	1	5	4	135	3	0	101	186	2	0	28	23	4	0	0	0	174	1691	BAS
NOS	3	2	19	37	5	0	20	1144	2	0	8	79	37	0	0	0	298	2965	NOS
ATL	45	3	183	73	8	0	113	1083	2	1	11	69	280	0	0	0	2148	5522	ATL
MED	14	44	233	6	36	35	69	85	62	34	2	9	13	1	0	0	897	3627	MED
BLS	0	46	1	4	1	25	281	9	7	12	1	1	0	0	0	0	205	877	BLS
SUM	188	577	1226	837	292	179	5159	5072	327	257	168	427	504	4	0	0	9578	49366	SUM
	PT	RO	ES	SE	CH	TR	SU	GB	YU	REM	BAS	NOS	ATL	MED	BLS	NAT	IND	SUM	

R e c e i v e r s

TABLE 4

Export/import budgets of oxidized sulphur and nitrogen for 1980 (Hundreds of tonnes of S and N)

	Oxidised Sulphur				Oxidised Nitrogen			
	Export mass (%)	Import mass (%)	% to sea	% in area	Export mass (%)	Import mass (%)	% to sea	% in area
AL	538 (90)	335 (84)	15	44	87 (96)	86 (96)	13	36
AT	1673 (84)	2030 (87)	8	60	705 (94)	706 (94)	10	52
BE	3601 (87)	1035 (66)	22	81	1301 (97)	267 (86)	25	77
BG	8950 (87)	1317 (50)	16	47	1181 (93)	283 (77)	13	40
CS	12765 (84)	4016 (62)	9	69	3184 (92)	952 (78)	12	65
DK	2060 (91)	655 (77)	35	78	813 (97)	240 (92)	28	72
FI	2233 (76)	1800 (72)	20	72	704 (88)	565 (85)	16	63
FR	12375 (74)	4503 (51)	23	74	4618 (83)	1681 (64)	25	69
DD	18138 (84)	2909 (46)	12	74	1463 (94)	978 (91)	16	71
DE	12050 (76)	5416 (59)	16	78	7845 (88)	1573 (60)	19	72
GR	1843 (92)	1184 (88)	24	40	884 (95)	227 (83)	18	33
HU	6682 (82)	1789 (55)	9	70	763 (92)	486 (88)	10	66
IS	27 (90)	100 (97)	40	47	38 (95)	57 (97)	33	33
IE	959 (86)	348 (70)	50	73	211 (95)	137 (93)	46	66
IT	16164 (85)	2327 (45)	28	54	3981 (88)	839 (62)	20	48
LU	113 (94)	102 (94)	13	79	69 (99)	30 (97)	19	74
NL	2127 (87)	1269 (80)	32	82	1716 (97)	294 (84)	29	76
NO	572 (81)	1644 (93)	31	68	513 (91)	674 (93)	24	54
PL	15178 (74)	7514 (59)	10	70	3972 (87)	1915 (76)	12	66
PT	1165 (88)	308 (65)	16	37	464 (92)	110 (73)	15	37
RO	6785 (77)	3307 (62)	9	58	986 (88)	773 (85)	10	51
ES	13431 (81)	1143 (27)	24	50	2450 (85)	549 (55)	18	42
SE	1956 (77)	2780 (83)	28	73	1136 (88)	1050 (87)	22	65
CH	516 (82)	717 (86)	8	61	562 (94)	270 (89)	10	49
TR	3343 (78)	2156 (69)	15	43	455 (86)	636 (89)	13	34
SU	34434 (56)	20580 (43)	7	57	6337 (63)	6843 (64)	7	51
GB	19349 (79)	1301 (20)	40	75	6663 (91)	478 (43)	40	70
YU	4918 (77)	3602 (71)	13	57	563 (88)	1038 (93)	13	51
RE	3488 (86)	3624 (86)	9	29	1000 (90)	1469 (93)	5	23
BAS	252 (70)	5876 (98)	37	77	215 (88)	1663 (98)	21	69
NOS	588 (68)	11525 (98)	45	80	505 (86)	2886 (97)	31	73
ATL	954 (60)	15964 (96)	43	55	781 (74)	5242 (95)	31	48
MED	58 (97)	14974 (100)	3	10	39 (97)	3626 (100)	3	10
BLS	- (-)	4323 (100)	-	-	- (-)	877 (100)	-	-

Mass in 100 tonnes Sulphur/Nitrogen, Export % of emission, Import % of depositions, % of emissions to sea, % of emissions retained in model area.

TABLE 5

Deposition of oxidized sulphur in 1993 (Hundreds of tonnes of S)

emitters

Receptor	AL	AT	BE	BG	DK	FI	FR	DE°	GR	HU	IS	IE	IT	LU	NL	NO	PL	PT	RO	ES	SE	CH	TR
AL	58	0	0	30	0	0	2	4	26	6	0	1	48	0	0	0	5	0	3	2	0	0	1 AL
AT	1	50	12	13	2	0	60	206	1	66	0	1	112	1	4	0	77	1	13	22	1	14	0 AT
BE	0	1	190	1	1	0	106	121	0	4	0	1	4	2	30	0	8	1	1	13	0	0	0 BE
BG	9	2	1	777	0	0	5	51	19	66	0	0	29	0	1	0	46	0	106	3	0	0	8 BG
DK	0	0	5	1	71	0*	8	101	0	5	0	2	1	0	5	1	55	0	1	2	4	0	0 DK
FI	0	1	4	1	8	124	6	66	0	9	0	1	0	0	2	4	82	0	2	1	19	0	0 FI
FR	1	6	146	5	3	0	1350	331	1	22	0	10	213	12	37	0	56	15	4	430	1	20	0 FR
DE°	1	21	225	15	34	0	435	4704	1	78	0	10	89	17	108	1	422	5	17	95	4	24	0 DE°
GR	34	1	1	308	0	0	5	22	190	20	0	0	63	0	0	0	18	0	22	3	0	0	16 GR
HU	2	15	5	30	1	0	18	137	4	646	0	0	56	0	2	0	135	0	51	8	0	0	1 HU
IS	0	0	0	0	0	0	1	3	0	0	3	1	0	0	0	0	1	0	0	0	0	2	1 IS
IE	0	0	2	0	0	0	9	10	0	0	0	99	0	0	0	0	2	1	0	14	0	0	0 IE
IT	5	10	7	35	1	0	143	112	13	56	0	1	1624	1	3	0	69	3	10	69	0	23	1 IT
LU	0	0	3	0	0	0	15	11	0	0	0	0	1	4	1	0	1	0	0	3	0	0	0 LU
NL	0	1	66	0	2	0	47	206	0	5	0	2	3	1	106	0	21	0	1	7	0	0	0 NL
NO	0	0	7	1	16	6	12	113	0	7	0	4	0	0	5	31	52	0	2	2	17	0	0 NO
PL	2	10	41	17	30	2	80	2169	2	164	0	4	38	2	23	1	3511	3	38	29	8	4	1 PL
PT	0	0	0	0	0	0	4	0	0	0	0	0	1	0	0	0	0	171	0	116	0	0	0 PT
RO	5	6	5	181	1	0	16	167	11	301	0	0	61	0	3	0	196	0	554	8	0	1	5 RO
ES	1	0	8	4	0	0	84	23	2	2	0	2	41	0	3	0	4	109	1	2064	0	1	0 ES
SE	0	1	14	2	71	25	22	260	0	19	0	4	2	0	10	19	199	0	8	3	107	0	0 SE
CH	0	2	5	2	0	0	68	38	0	4	0	0	94	1	2	0	7	1	1	24	0	49	0 CH
TR	7	2	3	242	1	0	10	69	82	42	0	0	49	0	1	0	63	0	50	4	0	0	407 TR
GB	0	1	39	0	4	0	89	186	0	7	0	50	4	1	25	0	44	2	1	29	1	0	0 GB
BY	0	3	6	20	10	3	16	213	1	41	0	1	12	0	4	1	387	0	20	11	5	1	1 BY
UA	3	8	14	147	9	3	36	488	10	258	0	2	59	1	8	1	891	1	167	14	4	2	17 UA
MD	0	0	1	12	0	0	2	26	1	16	0	0	5	0	0	0	41	0	23	0	0	0	1 MD
RU	3	7	21	142	39	111	49	662	16	177	0	5	57	1	14	1	905	1	112	23	38	3	45 RU
EE	0	0	1	0	3	10	2	32	0	2	0	0	1	0	1	1	31	0	1	1	4	0	0 EE
LV	0	0	2	2	6	3	4	61	0	4	0	0	1	0	1	1	70	0	2	1	5	0	0 LV
LT	0	0	3	5	7	1	5	85	0	8	0	0	2	0	2	0	125	0	5	2	4	0	0 LT
SI	0	6	1	3	0	0	7	21	1	14	0	0	60	0	0	0	15	0	3	4	0	1	0 SI
HR	2	5	2	13	1	0	11	53	2	63	0	0	115	0	1	0	49	0	8	7	0	1	1 HR
BA	4	3	1	19	0	0	7	40	4	51	0	0	75	0	1	0	37	0	7	6	0	0	1 BA
YU°	16	4	2	77	0	0	9	64	15	118	0	0	86	0	1	0	43	0	37	5	0	1	2 YU°
FYM	14	0	0	49	0	0	1	5	14	7	0	0	16	0	0	0	4	0	5	1	0	0	1 FYM
CS°	1	11	17	7	1	0	47	882	2	62	0	1	21	2	7	0	298	0	11	17	1	3	0 CS°
SK	1	8	4	9	1	0	12	132	2	156	0	0	18	0	2	0	192	0	18	5	0	0	1 SK
REM	2	1	6	47	2	2	41	78	14	21	0	1	148	0	3	0	82	5	18	78	1	2	18 REM
BAS	1	3	27	9	147	67	46	687	1	43	0	5	8	1	18	8	618	1	15	10	87	1	0 BAS
NOS	0	4	168	1	89	4	319	962	0	25	0	54	15	3	179	18	289	3	5	70	18	2	0 NOS
ATL	0	3	75	2	34	33	405	495	0	24	11	292	38	3	41	26	193	167	9	1750	25	2	0 ATL
MED	86	15	26	594	5	1	452	300	599	186	0	3	2882	2	11	0	214	27	92	725	2	16	195 MED
BLS	5	3	5	500	2	1	13	148	35	93	0	0	41	0	3	0	180	1	136	6	1	1	108 BLS
SUM	267	217	1172	3324	608	400	4079	14542	1071	2902	14	560	6192	60	670	124	9734	521	1581	5687	358	178	834 SUM
	AL	AT	BE	BG	DK	FI	FR	DE°	GR	HU	IS	IE	IT	LU	NL	NO	PL	PT	RO	ES	SE	CH	TR

emitters

Receptor	GB	BY	UA	MD	RU	EE	LV	LT	SI	HR	BA	YU°	FYM	CS°	SK	REM	BAS	NOS	ATL	MED	BLS	NAT	IND	SUM
AL	1	0	7	0	1	0	0	0	1	3	20	15	1	3	1	2	0	0	0	0	0	1	58	301 AL
AT	26	1	18	1	1	0	0	1	54	14	14	15	0	131	25	2	1	3	1	0	0	2	188	1157 AT
BE	87	0	3	0	0	0	0	0	1	0	0	1	0	17	1	0	0	14	2	0	0	2	42	653 BE
BG	5	3	122	7	9	1	0	1	3	9	34	88	2	30	15	2	0	1	0	0	0	2	191	1647 BG
DK	79	2	6	0	2	1	1	2	1	1	1	1	0	22	3	0	6	8	1	0	0	6	44	447 DK
FI	39	21	42	1	233	72	11	11	1	0	0	1	0	25	5	0	12	3	1	0	0	6	257	1071 FI
FR	337	1	9	0	1	0	0	1	11	8	10	8	0	72	9	10	1	48	33	0	0	26	614	3864 FR
DE°	463	4	52	2	8	3	1	4	17	9	9	13	0	877	36	1	10	56	10	0	0	15	578	8472 DE°
GR	2	2	51	2	0	0	0	2	6	27	39	2	0	12	4	4	0	0	1	0	0	3	189	1056 GR
HU	13	2	50	2	3	0	0	1	23	33	32	51	0	108	105	1	0	1	0	0	0	1	145	1688 HU
IS	7	0	0	0	0	0	0	0	0	0	0	0	0	1	0	0	0	1	0	0	0	10	23	53 IS
IE	119	0	1	0	0	0	0	0	0	0	0	0	0	2	0	0	0	2	9	0	0	13	53	339 IE
IT	26	1	21	1	2	0	0	0	46	43	64	29	0	50	15	24	0	2	2	0	0	9	516	3040 IT
LU	3	0	0	0	0	0	0	0	0	0	0	0	0	2	0	0	0	0	0	0	0	5	50	50 LU
NL	154	0	6	0	0	0	0	0	0	0	0	0	0	33	3	0	0	22	2	0	0	3	48	742 NL
NO	145	6	13	0	57	7	3	3	0	1	1	1	0	25	3	0	5	11	4	0	0	22	290	876 NO
PL	193	33	258	6	43	6	5	20	13	10	19	3	0	786	110	1	15	17	3	0	0	6	488	8238 PL
PT	4	0	0	0	0	0	0	0	0	0	0	0	0	0	0	2	0	19	0	0	0	6	57	382 PT
RO	16	8	358	25	18	1	1	3	12	24	47	118	1	108	77	2	1	2	0	0	0	2	358	2704 RO
ES	37	0	1	0	0	0	0	0	2	3	2	2	0	5	5	1	23	0	3	31	2	16	334	2814 ES
SE	168	19	48	2	75	24	10	15	1	1	2	4	0	70	10	0	25	16	4	0	0	19	418	1701 SE
CH	14	0	1	0	0	0	0	0	4	3	4	3	0	8	1	1	0	1	1	0	0	1	99	442 CH
TR	9	6	194	7	19	1	1	2	3	6	15	28	0	41	12	12	0	1	0	0	0	7	830	2228 TR
GB	3611	1	9	0	2	1	0	1	1	1	0	1	0	47	3	1	0	46	24	0	0	35	212	4483 GB
BY	42	392	309	6	98	13	15	47	4	3	5	2	0	89	21	1	6	3	1	0	0	2	299	2122 BY
UA	74	136	4930	75	234	11	8	22	15	19	24	43	0	232	115	11	5	6	1	0	0	5	921	9031 UA
MD	2	2	135	38	5	0	0	1	1	1	2	3	0	13	7	0	0	0	0	0	0	1	46	387 MD
RU	203	524	3182	39	6393	360	80	109	11	16	28	41	0	252	72	164	33	16	4	0	0	25	5232	19229 RU
EE	10	11	17	0	24	113	12	7	0	0	0	0	0	9	1	0	6	1	0	0	0	1	56	358 EE
LV	18	31	37	1	24	14	53	30	0	0	0	1	0	20	2	0	6	2	0	0	0	2	79	487 LV
LT	22	33	55	1	26	5	13	106	1	1	1	0	0	25	4	0	5	2	0	0	0	2	86	648 LT
SI	3	0	5	0	0	0	0	0	94	13	7	6	0	12	4	1	0	0	0	0	0	0	43	326 SI
HR	6	1	17	1	1	0	0	0	24	95	72	28	0	36	14	2	1	0	1	0	0	1	102	735 HR
BA	5	1	18	1	1	0	0	0	6	31	291	34	0	24	12	1	0	1	0	0	0	1	101	784 BA
YU°	7	1	38	2	2	0	0	1	7	26	195	374	2	41	20	4	0	1	0	0	0	1	167	1372 YU°
FYM	1	0	8	0	0	0	0	0	1	2	16	21	4	3	1	1	0	0	0	0	0	1	39	220 FYM
CS°	38	3	23	1	1	1	1	1	1	4	6	10	0	974	44	0	1	4	1	0	0	1	142	2659 CS°
SK	11	2	40	1	2	0	1	1	7	6	8	14	0	129	189	0	1	0	0	0	0	1	80	1055 SK
REM	23	26	517	5	332	6	2	5	3	4	10	10	0	22	6	588	1	2	2	1	0	13	1793	3941 REM
BAS	242	45	117	3	118	88	43	47	3	3	3	4	0	167	24	1	105	24	3	0	0	23	453	3325 BAS
NOS	3503	10	46	1	19	11	5	9	4	3	4	4	0	215	14	1	16	286	40	0	0	135	682	7235 NOS
ATL	1963	21	76	2	703	38	9	11	5	3	4	4	0	123	11	1	76	651	0	0	0	1408	4214	12965 ATL
MED	80	11	264	12	25	2	1	4	61	178	253	148	3	141	50	327	11	8	12	3	0	114	2304	10434 MED
BLS	17	20	905	33	88	3	2	1	6	12	31	48	0	129	30	29	1	2	0	0	0	15	777	3388 BLS
SUM	11829	1380	12009	279	8580	784	279	472	455	592	1264	1251	18	5084	1082	1227	277	693	868	7	0	1964	23659	129148 SUM
	GB	BY	UA	MD	RU	EE	LV	LT	SI	HR	BA	YU°	FYM	CS°	SK	REM	BAS	NOS	ATL	MED	BLS	NAT	IND	SUM

TABLE 6

Deposition of oxidized nitrogen in 1993 (Hundreds of tonnes of N)

emitters

Receptors	AL	AT	BE	BG	DK	FI	FR	DE*	GR	HU	IS	IE	IT	LU	NL	NO	PL	PT	RO	ES	SE	CH	TR
AL	4	1	0	3	0	0	3	3	8	1	0	0	28	0	0	0	2	0	2	1	0	0	1
AT	0	28	13	2	2	0	67	170	1	8	0	0	116	1	13	1	19	0	8	8	2	19	0
BE	0	1	36	0	1	0	46	78	0	1	0	0	3	1	30	0	2	1	0	5	1	1	0
BG	2	4	2	41	1	0	5	27	9	10	0	0	21	0	3	0	15	0	43	1	0	1	2
DK	0	1	4	0	23	0	7	43	0	1	0	2	1	0	13	3	18	0	1	1	7	0	0
FI	0	1	6	0	15	88	8	55	0	2	0	1	0	0	11	16	35	0	1	0	51	0	0
FR	0	11	89	1	4	0	730	315	1	4	0	7	165	6	76	1	18	17	3	180	2	24	0
DE*	0	31	149	2	30	2	367	1420	0	14	0	8	106	10	212	5	109	6	11	35	14	36	0
GR	5	2	1	22	0	0	6	12	44	3	0	0	36	0	1	0	6	0	12	2	1	1	5
HU	0	18	6	5	1	0	20	85	2	38	0	0	53	0	8	0	38	0	25	3	1	4	1
IS	0	0	1	0	0	0	1	2	0	0	1	0	0	0	1	0	0	0	0	0	1	0	0
IE	0	0	3	0	1	0	14	11	0	0	0	18	0	0	5	0	1	1	0	5	0	0	0
IT	1	13	8	5	2	0	151	92	6	9	0	1	743	1	10	1	20	2	7	36	2	25	1
LU	0	0	2	0	0	0	7	8	0	0	0	0	1	0	1	0	0	0	1	0	0	0	0
NL	0	1	23	0	3	0	28	99	0	1	0	1	2	1	61	1	6	0	1	3	1	0	0
NO	0	1	9	0	22	10	14	78	0	1	0	4	0	0	21	64	20	0	2	1	45	0	0
PL	0	19	43	3	36	3	89	580	1	25	0	4	42	2	75	7	492	1	21	9	25	10	0
PT	0	0	0	0	0	0	3	1	0	0	0	0	1	0	1	0	0	59	0	48	0	0	0
RO	1	11	6	21	2	0	17	84	5	40	0	0	48	0	9	1	62	0	127	3	2	2	2
ES	0	1	9	1	1	0	80	27	1	1	0	2	24	0	10	0	2	85	1	592	0	2	0
SE	0	3	16	1	60	39	25	149	0	4	0	4	2	0	37	59	69	0	7	1	138	1	0
CH	0	2	5	0	0	0	65	36	0	1	0	0	73	0	5	0	2	1	1	10	0	22	0
TR	1	4	3	24	1	1	11	39	36	7	0	0	25	0	3	0	23	0	31	2	1	1	81
GB	0	2	26	0	7	1	70	134	0	1	0	23	3	1	52	3	16	2	1	10	4	1	0
BY	0	4	7	3	15	5	19	98	1	7	0	1	11	0	13	4	104	2	11	3	17	3	0
UA	1	16	16	19	16	5	43	218	6	37	0	2	54	1	28	4	253	1	72	5	16	6	7
MD	0	1	1	1	0	0	2	11	0	2	0	0	3	0	1	0	12	0	7	0	1	0	0
RU	1	18	27	21	79	175	76	386	8	35	0	7	52	1	58	51	341	1	80	10	149	7	17
EE	0	1	1	0	4	8	3	16	0	0	0	0	0	0	3	3	10	0	1	0	10	0	0
LV	0	1	2	0	7	5	5	28	0	1	0	0	1	0	5	3	20	0	1	1	12	0	0
LT	0	1	3	1	9	2	6	37	0	1	0	0	2	0	7	2	32	0	3	1	12	1	0
SI	0	5	1	1	0	0	8	19	0	2	0	0	48	0	1	0	4	0	2	1	0	2	0
HR	0	7	2	2	1	0	12	35	1	7	0	0	89	0	3	0	14	0	5	3	1	2	0
BA	1	5	2	3	1	0	8	24	2	7	0	0	65	0	2	0	12	0	5	3	1	1	0
YU*	2	7	3	10	1	0	10	41	7	16	0	0	66	0	4	0	15	0	18	3	1	1	1
FYM	2	1	0	5	0	0	1	4	5	1	0	0	11	0	0	0	1	0	2	0	0	0	1
CS*	0	15	18	1	4	0	51	254	1	9	0	1	26	1	25	1	52	0	7	5	3	7	0
SK	0	9	4	2	1	0	14	65	1	15	0	0	19	0	7	0	40	0	9	2	1	3	0
REM	1	3	6	7	5	3	59	63	7	4	0	1	90	0	10	2	30	4	15	50	6	6	7
BAS	0	5	25	2	82	45	46	276	0	7	0	5	8	1	55	26	152	1	10	3	113	2	0
NOS	0	7	85	0	67	5	205	483	0	4	1	31	11	2	199	30	84	3	3	24	43	3	0
ATL	0	8	85	1	61	53	370	464	0	5	10	116	24	3	150	88	83	67	7	294	77	5	0
MED	12	26	24	51	6	1	376	225	150	28	0	2	1073	2	32	2	64	18	54	296	6	27	49
BLS	1	5	5	30	3	2	15	68	14	13	0	0	25	0	1	5	7	1	57	2	3	2	29
SUM	36	299	776	292	576	456	3166	6365	320	376	14	244	3174	41	1271	380	2360	273	672	1664	769	229	205

emitters

Receptors	GB	BY	UA	MD	RU	EE	LV	LT	SI	HR	BA	YU*	FYM	CS*	SK	REM	BAS	NOS	ATL	MED	BLS	NAT	IND	SUM
AL	1	0	2	0	1	0	0	0	0	1	2	0	0	0	0	0	0	0	0	0	0	0	19	87
AT	22	4	0	1	0	0	0	0	7	5	1	1	0	27	6	0	1	3	1	0	0	0	71	632
BE	55	0	1	0	0	0	0	0	0	0	0	0	0	5	1	0	7	2	0	0	0	0	19	298
BG	4	1	26	1	7	0	0	0	0	1	3	2	6	-11	6	1	0	1	0	0	0	0	57	316
DK	47	1	2	0	1	0	0	1	0	0	0	0	0	0	6	1	0	3	5	1	0	0	18	213
FI	37	9	13	0	47	13	6	4	0	0	0	0	0	11	2	0	10	4	1	0	0	0	99	551
FR	246	1	2	0	0	0	0	0	1	3	3	1	1	0	24	4	2	1	33	34	0	0	267	2276
DE*	321	2	12	0	2	0	0	0	2	4	4	1	1	0	162	16	0	6	41	12	0	0	250	3410
GR	2	1	12	1	4	0	0	0	0	0	2	2	3	0	5	2	1	0	0	0	0	0	61	258
HU	12	1	12	1	2	0	0	0	0	4	8	3	4	0	32	27	0	0	2	1	0	0	53	473
IS	6	0	0	0	0	0	0	0	0	0	0	0	0	0	0	0	0	0	1	0	0	0	19	36
IE	64	0	0	0	0	0	0	0	0	0	0	0	0	1	0	0	0	3	7	0	0	0	26	161
IT	21	1	6	0	0	0	0	0	7	12	6	3	0	15	6	4	0	3	3	0	0	0	190	1414
LU	2	0	0	0	0	0	0	0	0	0	0	0	0	0	0	0	0	0	0	0	0	0	2	29
NL	81	0	1	0	0	0	0	0	0	0	0	0	0	8	1	0	10	2	0	0	0	0	22	358
NO	107	3	5	0	7	2	2	2	0	0	0	0	0	9	2	0	5	10	3	0	0	0	122	576
PL	146	18	57	2	20	2	2	4	3	4	2	3	0	174	37	0	10	17	4	0	0	0	194	2187
PT	4	0	0	0	0	0	0	0	0	0	0	0	0	0	0	0	0	14	0	0	0	0	39	169
RO	14	3	63	5	14	0	0	0	1	3	8	4	8	0	37	30	1	1	2	1	0	0	112	750
ES	32	0	0	0	0	0	0	0	0	1	0	0	0	3	0	4	0	4	29	2	0	0	171	1087
SE	130	11	17	1	23	8	6	5	0	0	0	0	0	24	5	0	16	16	4	0	0	0	160	1041
CH	11	0	0	0	0	0	0	0	1	1	0	0	0	0	0	0	0	4	0	0	0	0	34	278
TR	8	3	48	2	26	0	0	0	1	2	1	3	0	16	6	7	0	1	0	0	0	0	287	707
GB	701	1	2	0	0	0	0	0	0	0	0	0	0	15	2	0	1	26	18	0	0	0	99	1224
BY	32	50	66	2	59	3	6	8	1	1	0	1	0	27	8	1	5	4	1	0	0	0	97	699
UA	64	43	489	13	150	3	4	7	4	7	2	4	0	76	42	8	5	7	2	0	0	0	314	2072
MD	2	1	21	2	4	0	0	0	0	0	0	0	0	4	3	0	0	0	0	0	0	0	14	97
RU	189	155	662	12	2083	54	38	36	3	7	3	4	0	104	37	91	35	21	6	0	0	0	1715	6858
EE	8	3	4	0	10	5	3	2	0	0	0	0	0	3	0	0	3	1	0	0	0	0	19	121
LV	13	7	9	0	11	3	5	4	0	0	0	0	0	6	1	0	3	2	0	0	0	0	27	184
LT	17	10	12	0	11	1	3	7	0	0	0	0	0	7	2	0	3	2	0	0	0	0	28	227
SI	2	0	1	0	0	0	0	0	5	1	1	0	0	3	1	0	0	0	0	0	0	0	16	129
HR	5	1	4	0	1	0	0	0	4	11	5	2	0	11	5	0	1	0	0	0	0	0	36	272
BA	4	1	4	0	1	0	0	0	2	8	3	3	0	8	5	0	1	0	0	0	0	0	34	223
YU*	6	1	8	0	2	0	0	0	2	9	8	14	0	15	8	1	1	0	0	0	0	0	56	338
FYM	1	0	2	0	1	0	0	0	0	1	1	2	0	2	1	0	0	0	0	0	0	0	12	56
CS*	35	1	6	0	0	0	0	0	2	2	1	1	0	105	13	0	1	5	2	0	0	0	61	719
SK	10	1	9	0	1	0	0	0	1	2	1	1	0	32	21	0	1	0	0	0	0	0	31	306
REM	21	10	120	2	210	2	2	1	2	2	1	2	0	8	4	116	2	1	0	0	0	0	921	1809
BAS	159	17	29	1	40	12	10	1	1	1	0	1	0	45	9	0	26	19	4	0	0	0	157	1405
NOS	1019	7	13	0	7	2	3	3	1	1	0	0	0	52	6	0	11	81	32	0	0	0	282	2812
ATL	1233	12	22	1	106	9	6	5	1	1	0	0	0	49	6	3	80	308	0	0	0	0	2310	6136
MED	61	5	64	3	22	0	1	1	11	31	19	14	0	43	21	37	2	8	11	2	0	0	925	3805
BLS	15	8	148	6	89	1	1	2	1	4	2	3	0	28	12	18	1	2	1	0	0	0	238	923
SUM	4968	389	1978	60	2974	123	103	109	77	150	79	89	2	1216	359	296	167	428	509	5	0	0	9684	47722

TABLE 7

Deposition of reduced nitrogen in 1993 (Hundreds of tonnes of N)

emitters

	AL	AT	BE	BG	DK	FI	FR	DE*	GR	HU	IS	IE	IT	LU	NL	NO	PL	PT	RO	ES	SE	CH	TR	
AL	95	1	0	4	0	0	1	1	5	1	0	0	5	0	0	0	1	0	2	0	0	0	1	AL
AT	0	393	1	4	1	0	35	160	0	21	0	0	54	1	6	0	10	0	10	2	0	23	0	AT
BE	0	1	264	0	1	0	61	28	0	1	0	1	1	2	55	0	1	0	1	2	0	0	0	BE
BG	3	3	0	753	1	0	2	8	10	15	0	0	4	0	1	0	7	0	113	0	0	0	6	BG
DK	0	1	1	0	437	1	0	3	30	0	1	0	1	0	0	6	1	11	0	1	0	5	0	DK
FI	0	1	1	0	6	166	2	11	0	2	0	1	0	0	2	3	19	0	1	0	9	0	0	FI
FR	0	6	68	1	2	0	2922	85	0	3	0	8	63	7	33	0	4	7	2	73	0	26	0	FR
DE*	0	45	81	2	40	0	241	3086	0	13	0	8	37	10	266	1	62	2	10	12	4	44	0	DE*
GR	13	1	0	45	0	0	1	3	236	4	0	0	6	0	0	0	2	0	15	0	0	0	10	GR
HU	1	29	1	9	0	0	7	29	1	477	0	0	12	0	3	0	16	0	49	1	0	2	1	HU
IS	0	0	0	0	0	0	1	1	0	0	10	1	0	0	0	0	0	0	0	0	0	0	0	IS
IE	0	0	0	0	0	0	9	3	0	0	0	533	0	0	1	0	1	0	0	2	0	0	0	IE
IT	1	16	2	5	1	0	66	46	3	11	0	0	1555	0	4	0	8	1	5	10	0	49	1	IT
LU	0	0	2	0	0	0	8	4	0	0	0	0	0	18	1	0	0	0	0	0	0	0	0	LU
NL	0	1	27	0	3	0	21	70	0	1	0	2	1	0	601	0	2	0	1	1	0	0	0	NL
NO	0	1	2	0	19	3	7	23	0	2	0	3	0	0	7	167	11	0	2	0	12	0	0	NO
PL	0	17	9	4	30	1	28	215	1	26	0	3	10	1	29	2	1659	0	25	2	9	4	1	PL
PT	0	0	0	0	0	0	2	0	0	0	0	0	0	0	0	0	0	336	0	54	0	0	0	PT
RO	1	8	1	73	1	0	5	20	2	86	0	0	10	0	3	0	25	0	1385	1	0	1	4	RO
ES	0	1	3	1	0	0	66	6	0	1	0	1	5	0	3	0	0	98	1	1338	0	1	0	ES
SE	0	2	3	1	70	13	9	49	0	4	0	3	0	0	11	27	41	0	7	0	247	0	0	SE
CH	0	3	2	1	0	0	63	34	0	1	0	0	47	0	2	0	1	0	1	4	0	250	0	CH
TR	2	3	1	37	0	0	4	11	12	8	0	0	5	0	1	0	9	0	33	1	0	0	1373	TR
GB	0	2	12	0	4	0	60	37	0	1	0	77	1	0	23	1	7	1	1	4	1	1	0	GB
BY	0	3	1	4	7	1	5	23	0	6	0	1	2	0	3	1	85	0	14	1	4	1	1	BY
UA	1	10	3	27	6	1	10	48	2	46	0	1	11	0	7	1	143	0	149	1	2	2	13	UA
MD	0	1	0	2	0	0	0	3	0	3	0	0	1	0	0	0	5	0	28	0	0	0	0	MD
RU	1	10	5	24	30	46	20	94	3	35	0	5	13	0	20	8	179	1	92	3	22	0	87	RU
EE	0	0	0	0	1	3	1	3	0	0	0	0	0	0	1	0	6	0	0	0	2	0	0	EE
LV	0	0	0	0	4	1	1	6	0	0	0	0	0	0	1	0	12	0	1	0	4	0	0	LV
LT	0	1	0	1	5	1	2	9	0	1	0	0	0	0	2	0	29	0	3	0	4	0	0	LT
SI	0	14	0	1	0	0	3	10	0	4	0	0	20	0	1	0	2	0	3	0	0	1	0	SI
HR	0	10	1	3	0	0	3	12	0	21	0	0	28	0	1	0	7	0	6	1	0	1	0	HR
BA	1	4	0	4	0	0	2	7	1	14	0	0	16	0	1	0	6	0	5	1	0	0	0	BA
YU*	6	5	0	29	0	0	2	10	4	33	0	0	13	0	1	0	6	0	30	1	0	0	1	YU*
FYM	6	0	0	13	0	0	0	1	6	1	0	0	2	0	0	0	0	0	3	0	0	0	1	FYM
CS*	0	34	5	2	2	0	21	139	0	12	0	1	7	1	11	0	39	0	7	2	1	4	0	CS*
SK	0	16	1	2	0	0	5	22	1	30	0	0	5	0	2	0	28	0	13	0	0	1	0	SK
REM	1	1	1	7	2	1	20	15	2	4	0	0	19	0	4	0	15	2	14	14	1	2	79	REM
BAS	0	4	5	2	147	24	15	137	0	6	0	4	2	0	20	7	119	0	8	1	70	1	0	BAS
NOS	0	5	55	0	129	1	264	205	0	4	0	37	3	1	153	21	39	1	3	8	14	1	0	NOS
ATL	0	5	24	1	31	12	402	112	0	4	6	283	6	1	46	36	40	61	6	138	11	2	0	ATL
MED	24	17	6	59	2	0	186	48	77	28	0	1	366	1	10	0	21	9	42	119	1	13	100	MED
BLS	1	3	1	0	4	16	5	14	0	0	0	5	0	0	3	0	21	0	91	1	0	1	175	BLS
SUM	160	675	596	1173	985	274	4591	4880	375	947	17	976	2337	45	1348	281	2702	522	2184	1797	427	436	1858	SUM
	AL	AT	BE	BG	DK	FI	FR	DE*	GR	HU	IS	IE	IT	LU	NL	NO	PL	PT	RO	ES	SE	CH	TR	

emitters

	GB	BY	UA	MD	RU	EE	LV	LT	SI	HR	BA	YU*	FYM	CS*	SK	REM	BAS	NOS	ATL	MED	BLS	NAT	IND	SUM	
AL	0	0	0	0	0	0	0	0	0	1	2	7	2	0	0	0	0	0	0	0	0	0	18	148	AL
AT	4	0	0	1	1	0	0	0	17	3	1	4	0	19	9	0	0	0	0	0	0	0	61	850	AT
BE	13	0	0	0	0	0	0	0	0	0	0	0	0	1	0	0	0	0	0	0	0	0	14	448	BE
BG	1	0	1	4	4	0	0	0	0	2	2	28	5	2	4	1	0	0	0	0	0	0	55	1035	BG
DK	9	0	0	0	0	0	0	0	1	0	0	0	0	1	1	0	0	0	0	0	0	0	12	526	DK
FI	4	0	0	0	0	27	13	10	10	0	0	0	0	2	1	0	0	0	0	0	0	0	72	366	FI
FR	58	0	0	0	0	0	0	0	1	1	1	1	0	3	1	2	0	0	0	0	0	0	178	3558	FR
DE*	60	0	0	1	2	0	1	2	4	2	1	2	0	36	8	0	0	0	0	0	0	0	189	4272	DE*
GR	0	0	0	1	3	0	0	0	0	1	2	11	7	1	1	1	0	0	0	0	0	0	52	420	GR
HU	1	0	0	1	1	0	0	0	6	10	3	15	0	11	31	0	0	0	0	0	0	0	42	761	HU
IS	1	0	0	0	0	0	0	0	0	0	0	0	0	0	0	0	0	0	0	0	0	0	7	22	IS
IE	42	0	0	0	0	0	0	0	0	0	0	0	0	0	0	0	0	0	0	0	0	0	16	607	IE
IT	3	0	0	0	1	0	0	0	7	10	6	7	0	3	3	4	0	0	0	0	0	0	154	1982	IT
LU	0	0	0	0	0	0	0	0	0	0	0	0	0	0	0	0	0	0	0	0	0	0	2	36	LU
NL	19	0	0	0	0	0	0	0	0	0	0	0	0	1	1	0	0	0	0	0	0	0	16	770	NL
NO	21	0	0	0	4	2	3	3	0	0	0	0	0	2	1	0	0	0	0	0	0	0	86	382	NO
PL	22	1	3	4	15	1	4	17	2	2	1	6	0	47	28	0	0	0	0	0	0	0	153	2383	PL
PT	1	0	0	0	0	0	0	0	0	0	0	0	0	0	0	0	0	0	0	0	0	0	15	409	PT
RO	2	0	3	26	9	0	1	1	2	6	4	24	1	8	16	1	0	0	0	0	0	0	100	1829	RO
ES	5	0	0	0	0	0	0	0	0	0	0	1	0	0	0	4	0	0	0	0	0	0	80	1617	ES
SE	20	0	0	1	1	14	6	10	13	0	0	1	0	5	3	0	0	0	0	0	0	0	122	684	SE
CH	2	0	0	0	0	0	0	0	0	0	1	0	1	0	0	0	0	0	0	0	0	0	33	448	CH
TR	1	0	2	4	14	0	1	1	1	1	1	5	0	1	3	10	0	0	0	0	0	0	253	1801	TR
GB	1689	0	0	0	0	0	0	0	0	0	0	0	0	0	0	0	0	0	0	0	0	0	61	1988	GB
BY	4	14	4	4	39	3	23	58	1	1	0	2	0	4	4	0	0	0	0	0	0	0	90	416	BY
UA	7	2	85	51	113	2	6	14	2	4	2	8	0	13	24	4	0	0	0	0	0	0	256	1077	UA
MD	0	0	0	140	2	0	0	0	0	0	0	1	0	1	1	0	0	0	0	0	0	0	12	203	MD
RU	28	5	23	27	4636	42	79	86	2	4	3	9	0	15	17	100	0	0	0	0	0	0	1573	7359	RU
EE	1	0	0	0	2	101	15	7	0	0	0	0	0	0	0	0	0	0	0	0	0	0	15	166	EE
LV	2	0	0	0	8	10	204	36	0	0	0	0	0	1	0	0	0	0	0	0	0	0	20	317	LV
LT	2	1	0	1	8	2	22	327	0	0	0	0	0	1	0	0	0	0	0	0	0	0	24	448	LT
SI	0	0	0	0	0	0	0	0	86	3	1	1	0	1	1	0	0	0	0	0	0	0	13	168	SI
HR	1	0	0	0	0	0	0	0	8	117	12	7	0	3	4	0	0	0	0	0	0	0	31	281	HR
BA	1	0	0	1	0	0	0	0	2	16	123	12	0	2	3	0	0	0	0	0	0	0	32	257	BA
YU*	1	0	0	0	1	0	0	0	1	8	15	359	6	3	5	1	0	0	0	0	0	0	48	592	YU*
FYM	0	0	0	0	0	0	0	0	1	0	0	12	52	0	0	0	0	0	0	0	0	0	12	115	FYM
CS*	5	0	0	1	1	0	1	1	2	1	1	2	0	274	11	0	0	0	0	0	0	0	45	633	CS*
SK	1	0	0	1	1	0	0	0	2	1	1	3	0	14	196	0	0	0	0	0	0	0	24	374	SK
REM	2	0	3	3	198	1	2	3	1	1	0	3	1	3	1	468	0	0	0	0	0	0	498	1391	REM
BAS	25	1	1	1	23	25	38	37	1	1	0	2	0	7	5	0	0	0	0	0	0	0	119	860	BAS
NOS	398	0	0	1	3	2	3	5	1	1	0	1	0	9	5	0	0	0	0	0	0	0	182	1556	NOS
ATL	409	0	0	1	39	4	6	7	1	1	0	1	0	8	3	1	0	0	0	0	0	0	1281	2989	ATL
MED	7	0	2	5	9	0	1	2	9	27	19	29	5	7	8	39	0	0	0	0	0	0	590	1887	MED
BLS	2	0	7	16	72	0	1	3	1	2	2	7	1	5	5	18	0	0	0	0	0	0	213	749	BLS
SUM	2873	26	141	301	5263	217	431	639	161	230	206	572	83	519	403	656	0	0	0	0	0	0	6871	49178	SUM
	GB	BY	UA	MD	RU	EE	LV	LT	SI	HR	BA	YU*	FYM	CS*	SK	REM	BAS	NOS	ATL	MED	BLS	NAT	IND	SUM	

(R e c e i v e r s)

TABLE 8

Export/import budgets of oxidized sulphur and nitrogen for 1993 (Hundred of tonnes of S and N)

	Oxidised Sulphur				Oxidised Nitrogen				Reduced Nitrogen			
	Export mass (%)	Import mass (%)	% to sea	% in area	Export mass (%)	Import mass (%)	% to sea	% in area	Export mass (%)	Import mass (%)	% to sea	% in area
AL	542 (90)	243 (81)	17	45	87 (96)	83 (95)	14	39	152 (62)	53 (36)	10	65
AT	305 (86)	1107 (96)	8	61	526 (95)	604 (96)	9	54	439 (53)	457 (54)	4	81
BE	1330 (87)	463 (71)	22	77	1029 (97)	262 (88)	21	73	370 (58)	184 (41)	14	94
BG	6333 (89)	370 (53)	18	47	683 (94)	275 (87)	12	40	1058 (58)	282 (27)	6	65
DK	714 (91)	376 (84)	38	77	780 (97)	190 (89)	27	72	600 (58)	89 (17)	30	95
FI	481 (79)	947 (88)	20	66	682 (89)	463 (84)	14	59	171 (51)	200 (55)	11	81
FR	4329 (76)	2514 (65)	25	72	3893 (84)	1546 (68)	22	68	2224 (43)	636 (18)	17	89
DE*	14775 (76)	3768 (44)	15	75	7418 (84)	1990 (58)	17	72	2159 (41)	1186 (28)	10	93
GR	2360 (93)	866 (82)	29	42	887 (95)	214 (83)	18	34	406 (63)	184 (44)	13	58
HU	3489 (84)	1042 (62)	10	70	519 (93)	435 (92)	10	68	676 (59)	284 (37)	5	82
IS	27 (90)	50 (94)	43	47	35 (97)	35 (97)	30	38	14 (59)	12 (55)	24	69
IE	686 (87)	240 (71)	54	71	353 (95)	143 (89)	41	66	504 (49)	74 (12)	31	94
IT	9630 (86)	1416 (47)	30	55	5505 (88)	671 (47)	18	51	1607 (51)	427 (22)	12	74
LU	76 (95)	46 (92)	11	75	58 (100)	29 (100)	14	71	31 (64)	18 (50)	6	91
NL	734 (87)	636 (86)	33	80	1646 (96)	297 (83)	26	74	807 (57)	169 (22)	16	96
NO	154 (83)	845 (96)	31	67	620 (91)	512 (89)	21	56	162 (49)	215 (56)	19	86
PL	10114 (74)	4727 (57)	12	71	2977 (86)	1695 (78)	13	68	1486 (47)	724 (30)	8	86
PT	1279 (88)	211 (55)	16	36	687 (92)	110 (65)	12	37	430 (56)	73 (18)	9	68
RO	2241 (80)	2150 (80)	10	57	1221 (91)	623 (83)	10	50	1670 (55)	444 (24)	5	71
ES	9515 (82)	750 (27)	26	49	3233 (85)	495 (46)	16	44	1329 (50)	279 (17)	10	67
SE	398 (79)	1594 (94)	29	71	1076 (89)	903 (87)	20	63	230 (48)	437 (64)	20	90
CH	241 (83)	393 (89)	8	61	434 (95)	256 (92)	9	50	269 (52)	198 (44)	3	84
TR	1363 (77)	1821 (82)	20	47	451 (85)	626 (89)	15	39	2043 (60)	428 (24)	8	54
GB	12359 (77)	872 (19)	41	74	6466 (90)	523 (43)	35	69	1456 (46)	299 (15)	27	91
BY	1773 (82)	1730 (82)	5	64	580 (92)	649 (93)	8	62	18 (57)	402 (97)	3	80
UA	14319 (74)	4101 (45)	8	62	2849 (85)	1583 (76)	8	59	103 (55)	992 (92)	6	75
MD	417 (92)	349 (90)	13	61	104 (98)	95 (98)	10	56	272 (66)	63 (31)	6	73
RU	10882 (63)	12836 (67)	7	50	4820 (70)	4775 (70)	4	43	2793 (38)	2723 (37)	2	71
EE	1087 (91)	245 (68)	13	65	196 (98)	116 (96)	12	61	154 (60)	65 (39)	12	85
LV	357 (87)	434 (89)	16	68	159 (97)	179 (97)	13	63	298 (59)	113 (36)	10	86
LT	574 (84)	542 (84)	12	69	163 (96)	220 (97)	12	64	406 (55)	121 (27)	7	87
SI	816 (90)	232 (71)	10	50	168 (97)	124 (96)	9	44	136 (61)	82 (49)	6	72
HR	805 (89)	640 (87)	25	66	241 (96)	261 (96)	15	59	188 (62)	164 (58)	11	76
BA	2109 (88)	493 (63)	14	53	156 (95)	215 (96)	13	48	173 (58)	134 (52)	7	70
YU*	1631 (81)	998 (73)	12	62	150 (91)	324 (96)	11	54	456 (56)	233 (39)	5	70
FYM	46 (92)	216 (98)	6	36	6 (100)	56 (100)	0	33	96 (65)	63 (55)	4	56
CS*	6121 (86)	1685 (63)	11	72	1642 (94)	614 (85)	12	70	335 (55)	359 (57)	6	85
SK	1436 (88)	866 (82)	9	67	539 (96)	285 (93)	10	64	306 (61)	178 (48)	5	80
RE	3479 (86)	3353 (85)	10	30	991 (90)	1693 (94)	5	27	931 (67)	923 (66)	4	47
BAS	255 (71)	3220 (97)	38	77	217 (89)	1379 (98)	22	69	0 (0)	860 (100)	-	-
NOS	584 (67)	6949 (96)	46	80	503 (86)	2731 (97)	33	73	0 (0)	1556 (100)	-	-
ATL	927 (59)	12314 (95)	45	55	753 (71)	5828 (95)	34	48	0 (0)	2989 (100)	-	-
MED	57 (95)	10431 (100)	5	12	38 (95)	3803 (100)	5	13	0 (0)	1887 (100)	-	-
BLS	0 (0)	3388 (100)	-	-	0 (0)	923 (100)	-	-	0 (0)	749 (100)	-	-

Mass in 100 tonnes Sulphur/Nitrogen, Export % of emission, Import % of depositions, % of emissions to sea, % of emissions retained in model area.

TABLE 9

**Selected results from the time series analysis of monitoring results—
Sulphur dioxide. Median annual reductions, and concentration reductions
from 1980 to 1993 in per cent of the 1980/81 concentrations**

Site	Median annual reduction µg S/m3.year	Concentration change 1980-1993 (per cent of 1980/1981)
Rorvik, SE2	-0.22	-72
Bredkalen, SE5	-0.081	-81
Hoburg, SE8	-0.13	-64
Suwalki, PL1	not significant	-
Westerland, DE1	-0.21	-
Langenbrugge, DE2	-0.60	-71
Deuselbach, DE4	-0.55	-85
Schauinsland, DE3	-0.15	-
Brotjacklriegel, DE5	-0.27	-71
Valentia Obs., IE1	not significant	-
Eskdalemuir, GB2	-0.20	-56
Birkenes, NO1	-0.032	-
Skreådalen, NO8	-0.039	-50
Tustervatn, NO15	-0.037	-73
Kårvatn, NO39	-0.031	-75
Tange, DK3	-0.28	-76
Keldsnor	-0.31	-67

TABLE 10

**Selected results from the time series analysis of monitoring results—
Sulphate in particles. Median annual reduction, and concentration reduction
from 1980 to 1993 in per cent of the 1980/81 concentrations**

Site	Median annual reduction µg S/m3.year	Concentration change 1980-1993 (per cent of 1980/1981)
Ansbach, DE17	-0.067	-43
Brotjacklriegel, DE5	not significant	-
Rottenburg, DE18	not significant	-
Starnberg, DE19	not significant	-
Payerne, CH2	not significant	-
Valentia Obs. IE1	not significant	-
Eskdalemuir, GB2	not significant	-
Birkenes, NO1	-0.036	-
Skreådalen, NO8	-0.034	-43
Tustervatn, NO15	-0.027	-58
Kårvatn, NO39	-0.018	-46
Tange, DK3	-0.085	-43
Keldsnor, DK5	-0.091	-38
Rorvik, SE2	-0.080	-50
Hoburg, SE8	-0.012	-12
Ahtari, F14	-0.034	-
Westerland, DE1	not significant	-
Langenbrugge, DE2	not significant	-
Bassum, DE12	-0.073	-40
Meinerzhagen, DE14	-0.048	-31
Deuselbach, DE4	-0.067	-

FIGURE 1

Total deposition of oxidized sulphur in 1980 (cg(S)/m^2/year)

Row	6	7	8	9	10	11	12	13	14	15	16	17	18	19	20	21	22	23	24	25	26	27	28	29	30	31	32	33	34	35	36	37	38
36	3	3	4	4	5	6	8	15	18	19	23	25	26	28	32	37	42	47	56	63	93	73	88	85	74	56	40	31	24	20	15	13	10
35	3	3	4	4	6	8	10	12	15	20	23	24	27	30	33	39	45	49	55	66	86	85	75	64	57	47	41	34	23	20	17	15	20
34	3	3	4	5	7	8	10	13	18	25	27	27	27	34	38	44	50	57	61	70	75	85	92	79	61	46	41	33	30	30	25	24	20
33	3	3	5	6	7	8	11	15	22	29	33	33	33	49	47	49	55	64	66	71	81	92	113	144	65	52	44	33	28	34	26	23	18
32	3	4	5	5	7	8	12	19	26	36	43	49	43	48	78	55	59	67	73	81	94	91	81	76	72	55	50	48	48	31	27	24	19
31	3	4	5	6	4	7	12	19	29	48	80	90	61	52	60	57	65	73	86	104	146	99	89	73	79	60	55	57	35	27	27	29	18
30	3	4	5	7	8	9	14	21	32	41	205	150	39	52	69	84	77	85	102	140	120	124	111	90	75	74	81	63	45	40	48	39	22
29	3	3	5	7	10	11	16	20	24	25	47	61	60	94	79	115	103	104	113	127	206	125	101	77	95	97	99	75	51	44	47	37	24
28	3	3	5	7	9	12	15	19	25	28	41	64	64	72	84	96	172	128	101	127	196	218	113	100	94	188	199	74	55	50	40	33	19
27	3	4	6	7	10	13	15	20	27	32	40	55	63	89	92	124	149	120	119	122	138	136	136	144	154	390	215	83	55	47	45	40	18
26	4	5	5	7	11	14	15	19	33	35	39	60	88	87	114	149	172	118	123	157	165	141	150	142	201	224	164	90	63	36	37	39	24
25	5	5	6	8	12	14	16	20	28	37	37	49	77	96	122	143	124	131	155	204	181	189	159	156	199	333	128	115	62	44	41	52	21
24	6	6	7	9	12	14	18	22	29	40	39	45	71	88	95	104	158	154	162	188	157	169	161	227	168	288	148	95	63	51	59	55	23
23	7	7	8	10	13	16	20	25	33	49	36	50	74	93	139	130	154	169	191	194	187	209	170	201	162	147	145	136	72	115	76	60	28
22	6	7	9	11	13	17	20	29	41	47	37	69	103	116	115	162	176	222	231	252	255	209	209	186	162	253	134	105	87	93	72	50	24
21	7	8	15	12	14	18	24	33	45	77	70	82	131	149	168	195	218	263	303	391	336	275	313	209	182	182	116	134	99	83	52	101	25
20	8	7	15	15	15	18	26	37	50	79	122	138	172	193	250	232	320	319	409	512	460	345	285	258	205	253	196	312	130	223	82	49	27
19	14	13	16	16	17	20	26	40	58	75	101	178	205	181	235	354	434	465	682	444	461	294	216	204	276	242	235	149	123	79	44	36	
18	11	13	14	15	18	21	28	41	59	82	102	121	158	197	294	310	524	999	729	464	424	440	251	276	229	214	215	402	210	78	62	41	61
17	11	13	13	15	18	22	31	42	77	126	124	146	178	223	313	388	736	969	492	347	395	382	243	298	224	171	199	163	127	101	102	97	51
16	11	12	13	15	18	23	32	71	97	376	255	221	220	214	472	500	425	344	300	313	299	243	225	273	158	120	112	107	118	98	73	54	35
15	12	11	13	14	17	24	33	60	99	113	544	805	311	513	768	485	394	277	321	228	224	247	171	178	158	153	118	87	86	123	65	71	86
14	11	12	13	15	17	23	33	56	116	152	813	424	563	569	463	366	291	228	227	171	247	233	153	139	124	158	306	99	90	90	73	53	33
13	11	12	12	14	17	21	35	56	78	120	207	274	235	228	274	206	196	211	252	231	183	157	122	95	94	176	94	74	60	53	47	39	29
12	11	11	11	13	16	20	32	85	67	90	116	155	216	311	224	145	181	158	200	226	201	193	130	148	123	129	107	71	54	45	41	35	39
11	10	11	12	13	16	19	25	38	51	65	97	128	115	136	130	124	170	171	137	168	98	100	84	82	118	279	162	75	55	41	36	43	21
10	10	11	12	14	17	19	25	35	41	53	73	79	96	147	114	123	114	141	187	120	68	69	69	129	338	126	78	51	36	31	28	17	
9	10	10	12	14	16	20	23	30	35	45	60	68	80	92	104	103	108	96	113	81	89	91	86	68	105	71	68	53	39	31	27	24	18
8	10	10	12	13	15	17	19	26	32	39	47	55	72	88	95	125	82	112	95	70	65	86	66	68	58	46	41	36	31	25	23	24	16
7	11	11	12	12	14	15	18	23	28	40	54	108	202	164	134	101	91	100	84	62	52	51	53	46	36	34	36	31	24	15	13	12	10
6	9	10	10	11	12	15	17	22	28	46	119	253	267	121	70	66	198	201	80	49	46	70	45	28	26	31	32	29	25	12	10	10	9
5	9	9	10	11	12	14	17	21	31	48	132	104	58	49	75	52	68	82	47	43	64	39	19	19	15	15	16	13	11	9	8	8	8
4	9	9	9	10	11	14	15	19	26	41	65	82	49	34	44	45	45	64	43	53	41	23	14	14	9	10	9	9	9	8	7	7	8
3	8	8	9	9	10	12	15	19	23	31	44	65	52	37	43	49	30	87	45	33	20	15	25	11	9	8	7	8	7	7	7	7	7
2	8	8	8	9	10	11	14	16	20	26	33	53	88	37	36	37	32	38	29	20	16	17	11	10	13	8	7	7	7	7	7	7	7

FIGURE 2

Total deposition of oxidized sulphur in 1990 (cg(S)/m^2/year)

FIGURE 3

Change in exceedance of the critical sulphur deposition 1980-1990
(Percentage reduction in exceedance)

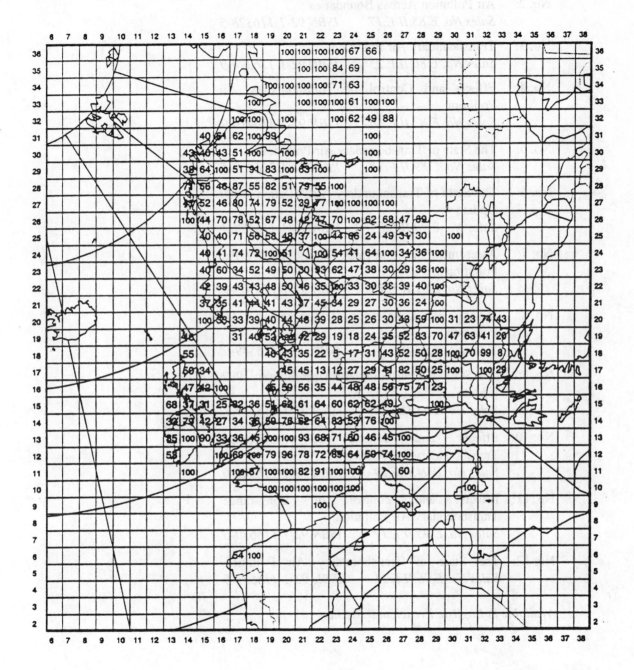

In this series:

No. 1: Airborne Sulphur Pollution
Sales No. E.84.II.E.8   ISBN 92-1-116307-2

No. 2: Air Pollution Across Boundaries
Sales No. E.85.II.E.17   ISBN 92-1-116328-5

No. 3: Transboundary Air Pollution
Sales No. E.86.II.E.23   ISBN 92-1-116374-9

No. 4: Effects and Control of Transboundary Air Pollution
Sales No. E.87.II.E.36   ISBN 92-1-116410-9

No. 5: The State of Transboundary Air Pollution
Sales No. E.89.II.E.25   ISBN 92-1-116460-5

No. 6: The State of Transboundary Air Pollution: 1989 Update
Sales No. E.90.II.E.33   ISBN 92-1-116489-3

No. 7: Assessment of Long-range Transboundary Air Pollution
Sales No. E.91.II.E.18   ISBN 92-1-116505-9

No. 8: Impacts of Long-range Transboundary Air Pollution
Sales No. E.92.II.E.24   ISBN 92-1-116549-0

No. 9: The State of Transboundary Air Pollution: 1992 Update
Sales No. E.93.II.E.25   ISBN 92-1-116574-1

No. 10: Effects and control of Long-range Transboundary Air Pollution
Sales No. E.94.II.E.24   ISBN 92-1-116602-0

No. 11: Effects and control of Long-range Transboundary Air Pollution
Sales No. E.95.II.E.17   ISBN 92-1-116630-6

No. 12: The State of Transboundary Air Pollution
Sales No. E.96.II.E.21   ISBN 92-1-116653-5

Available in English, French and Russian